AF581767

Biomarkers of Stress in Companion Animals

Biomarkers of Stress in Companion Animals

Editors

Angelo Gazzano
Asahi Ogi

MDPI • Basel • Beijing • Wuhan • Barcelona • Belgrade • Manchester • Tokyo • Cluj • Tianjin

Editors

Angelo Gazzano
University of Pisa
Pisa
Italy

Asahi Ogi
University of Pisa
Pisa
Italy

Editorial Office
MDPI
St. Alban-Anlage 66
4052 Basel, Switzerland

This is a reprint of articles from the Special Issue published online in the open access journal *Animals* (ISSN 2076-2615) (available at: https://www.mdpi.com/journal/animals/special_issues/biomarker_stress_companion_animals).

For citation purposes, cite each article independently as indicated on the article page online and as indicated below:

LastName, A.A.; LastName, B.B.; LastName, C.C. Article Title. *Journal Name* **Year**, *Volume Number*, Page Range.

ISBN 978-3-0365-7124-9 (Hbk)
ISBN 978-3-0365-7125-6 (PDF)

Contents

About the Editors

Angelo Gazzano

Doctor in Veterinary Medicine, he received a PhD degree in veterinary physiology from the University of Pisa (Italy). Angelo Gazzano is a diplomate of the European College of Animal Welfare and Behavioural Medicine, and European Veterinary Specialist (EBVS®) in Animal Welfare Science Ethics and Law, and Behavioural Medicine. He is currently Associate Professor at Department of Veterinary Sciences (University of Pisa) and head of ETOVET - Laboratory in Veterinary Ethology and Physiology (University of Pisa). His research is mainly focused on animal behavior and welfare, veterinary physiology, applied ethology, anthrozoology, behavioral medicine, and dog training. Dr. Gazzano is also designated veterinarian with expertise in laboratory animal medicine (Directive 2010/63/EU) at University of Pisa.

Asahi Ogi

Doctor in Veterinary Medicine, he received a PhD degree in veterinary physiology from the University of Pisa (Italy). Asahi Ogi is a diplomate of the European College of Animal Welfare and Behavioural Medicine, and European Veterinary Specialist (EBVS®) in Behavioural Medicine. He is currently Adjunct Professor at Department of Veterinary Sciences (University of Pisa) and member of ETOVET - Laboratory in Veterinary Ethology and Physiology (University of Pisa). His research is mainly focused on animal behavior and welfare, veterinary physiology, applied ethology, anthrozoology, behavioral medicine, and dog training. Dr. Ogi is also designated veterinarian with expertise in laboratory animal medicine (Directive 2010/63/EU) at University of Pisa and postdoctoral researcher at IRCCS Stella Maris Foundation - Molecular Medicine for Neurodegenerative and Neuromuscular Diseases Unit (Pisa).

 animals MDPI

Editorial

Biomarkers of Stress in Companion Animals

Asahi Ogi * and Angelo Gazzano

Department of Veterinary Science, University of Pisa, 56124 Pisa, Italy
* Correspondence: a.ogi@hotmail.com

1. Introduction

Stress experienced by companion animals could impair their physical and psychological welfare, impacting their social relationships in domestic environments [1]. For this reason, the establishment of possible reliable biomarkers—objective, quantifiable characteristics of biological processes [2]—is traditionally a goal of scientific research in veterinary behavioral medicine and animal welfare.

The aim of this book was to collect new insight and knowledge on biomarkers as a "characteristic that is measured as an indicator of normal biological processes, pathogenic processes, or biological responses to an exposure or intervention, including therapeutic interventions" [3].

2. Cortisol and Dehydroepiandrosterone Sulfate as Possible Biomarkers of Stress during Maternity of Dogs

Cortisol, dehydroepiandrosterone (DHEA) and dehydroepiandrosterone sulfate (DHEA-S) are final products of the activation of the hypothalamic–pituitary–adrenal (HPA) axis. Most previous studies relate to these biomarkers in blood, urine, feces and saliva, but none of these matrices are suitable when long-lasting physiological processes have to be investigated [4]. The use of different matrices, such as claws and hair, has been proposed to overcome these issues, because they provide retrospective information about long-term hormone accumulation [4]. Despite the fact that the oxytocinergic system seems to be triggered by mother–infant interactions [5] and, in turn, oxytocin release could exert inhibitory action on the HPA axis [6], maternity seems to play a crucial role in HPA axis activation. Indeed, maternity in dogs leads to the long-term accumulation of cortisol, but not of DHEA-S, in the coats and nails of female dogs during pregnancy and postpartum [4].

Citation: Ogi, A.; Gazzano, A. Biomarkers of Stress in Companion Animals. *Animals* **2023**, *13*, 660. https://doi.org/10.3390/ani13040660

Received: 3 January 2023
Revised: 7 February 2023
Accepted: 7 February 2023
Published: 14 February 2023

3. The Weight of Parturient and Blood Gas Analysis as Possible Biomarkers of Intrauterine Asphyxia in Newborn Canines

Intrapartum hypoxia/asphyxia negatively impacts newborn puppies' adaptation to extrauterine life [7]. Blood gas analysis allows us to estimate variations in perinatal oxygenation levels, metabolic profiles and the acid–base balance [8]. During eutocic birth, alterations in gases and blood metabolites could indicate respiratory and metabolic acidosis resulting from intrapartum asphyxia [8]. Furthermore, the weight of dams determines the weight of the puppies at birth and, consequently, predicts possible difficulties during expulsion through the birth channel. Indeed, puppies with higher weights have a higher risk of suffering an acute process of intrauterine asphyxia [8].

4. Blood Oxytocin, Prolactin, and Serotonin as Possible Biomarkers of Different Lifestyles in Dogs

Oxytocin release in dogs has often been associated with positive experience and, in particular, to positive human–dog interaction [9]. Furthermore, comparing the oxytocinergic system of dogs and wolves, life experience seems to be even more influential on oxytocin levels than domestication [10]. Similarly, the oxytocinergic system of assistance dogs could be more representative of their stressful lifestyle—shaped by repeated separation from their foster family—than their acknowledged prosociality [11]. In fact, even though oxytocin has been frequently linked with more friendliness or sociability, free and total blood oxytocin were found to be surprisingly lower in assistance dogs than pet dogs [11].

Increased levels of prolactin have been previously linked to acute and chronic stress in dogs, but results are conflicting, conceivably because this neurohormone seems to be strongly related to individual variability [11]. In comparison with prolactin, serotonin seems to be a more reliable biomarker. Indeed, low levels of serotonin were found to be associated with stress and behavioral disorders in dogs [12]. Moreover, L-tryptophan (serotonin precursor) [13] and serotonin reuptake inhibitors [14,15] are drugs widely used to treat behavioral problems. However, a clear association of this monoamine with lifestyle in dogs was not found [11], possibly because peripheral serotonin also seems to be affected by dietary patterns [16].

5. Facial Expressions as Possible Biomarkers of Emotions in Dogs

Facial expressions, a form of nonverbal communication, not only convey positive and negative emotions, but also have the purpose of contributing to calming/de-escalating other interacting dogs [17]. Despite the fact that the literature regarding possible tools for standardizing the Facial Action Coding System (FACS) is quite extensive, the association with underlying emotions is still difficult to evaluate on an objective basis [18]. The inter-species and the interbreed variability of facial mimicry, in fact, makes the standardization very complex [18]. However, through the FACS, some veterinary tools (e.g., the Grimace Scale) are currently available to help clinicians recognize pain and distress in various domestic animals [18]. These tools are crucial in improving the interpretation of animal body language and therefore the attitude of veterinarians towards animal welfare [19].

6. Skin Temperature as Possible Biomarker of Stress in Companion Animals

Stressful events are known to trigger sympathetic activation which, in turn, could cause an increase in body temperature. Infrared thermography (IRT) exploits a specific camera to detect the radiation spectrum and visualize changes in skin temperature which can be correlated with possible health/stress conditions in companion animals [20]. IRT is a safe and non-stressful technique, but currently, it is difficult to say that the evidence of its reliability is conclusive, mostly because of the possible influence of environmental factors [20].

7. Conclusions

Oxytocin, cortisol, DHEA-S, prolactin, serotonin, facial expressions, skin temperature, and birthweight are just some of possible biomarkers of stress in companion animals. In addition, all of these parameters could also be influenced by many other variables, such as the traits of the subjects (age, sex, breed, temperament, etc.) and methodological approaches (matrix, assay, instrumentation, etc.). In light of this complex picture, the present topic needs and deserves further investigations to optimally manage the stress experienced by companion animals and improve their welfare.

Author Contributions: Conceptualization, A.O.; writing—original draft preparation, A.O.; writing—review and editing, A.O. and A.G. All authors have read and agreed to the published version of the manuscript.

Funding: This research received no external funding.

Conflicts of Interest: The authors declare no conflict of interest.

References

1. Neilson, J.C. Fear of places or things. In *BSAVA Manual of Canine and Feline Behavioural Medicine*; British Small Animal Veterinary Association: Gloucester, UK, 2002; pp. 173–180.
2. Strimbu, K.; Tavel, J.A. What are biomarkers? *Curr. Opin. HIV AIDS* **2010**, *5*, 463–466. [CrossRef] [PubMed]

3. FDA-NIH Biomarker Working Group. *BEST (Biomarkers, EndpointS, and other Tools) Resource*; Food and Drug Administration (US): Silver Spring, MD, USA; National Institutes of Health (US): Bethesda, MD, USA, 2021.
4. Fusi, J.; Peric, T.; Probo, M.; Cotticelli, A.; Faustini, M.; Veronesi, M.C. How Stressful Is Maternity? Study about Cortisol and Dehydroepiandrosterone-Sulfate Coat and Claws Concentrations in Female Dogs from Mating to 60 Days Post-Partum. *Animals* **2021**, *11*, 1632. [CrossRef] [PubMed]
5. Ogi, A.; Naef, V.; Santorelli, F.M.F.M.; Mariti, C.; Gazzano, A. Oxytocin Receptor Gene Polymorphism in Lactating Dogs. *Animals* **2021**, *11*, 3099. [CrossRef] [PubMed]
6. Ogi, A.; Mariti, C.; Pirrone, F.; Baragli, P.; Gazzano, A. The Influence of Oxytocin on Maternal Care in Lactating Dogs. *Animals* **2021**, *11*, 1130. [CrossRef] [PubMed]
7. Vassalo, F.G.; Simões, C.R.B.; Sudano, M.J.; Prestes, N.C.; Lopes, M.D.; Chiacchio, S.B.; Lourenço, M.L.G. Topics in the Routine Assessment of Newborn Puppy Viability. *Top. Companion Anim. Med.* **2015**, *30*, 16–21. [CrossRef] [PubMed]
8. Reyes-Sotelo, B.; Mota-Rojas, D.; Mora-Medina, P.; Ogi, A.; Mariti, C.; Olmos-Hernández, A.; Martínez-Burnes, J.; Hernández-ávalos, I.; Sánchez-Millán, J.; Gazzano, A. Blood biomarker profile alterations in newborn canines: Effect of the mother's weight. *Animals* **2021**, *11*, 2307. [CrossRef] [PubMed]
9. Ogi, A.; Mariti, C.; Baragli, P.; Sergi, V.; Gazzano, A. Effects of Stroking on Salivary Oxytocin and Cortisol in Guide Dogs: Preliminary Results. *Animals* **2020**, *10*, 708. [CrossRef] [PubMed]
10. Wirobski, G.; Range, F.; Schaebs, F.S.; Palme, R.; Deschner, T.; Marshall-Pescini, S. Life experience rather than domestication accounts for dogs' increased oxytocin release during social contact with humans. *Sci. Rep.* **2021**, *11*, 14423. [CrossRef] [PubMed]
11. Mengoli, M.; Oliva, J.L.; Mendonça, T.; Chabaud, C.; Arroub, S.; Lafont-Lecuelle, C.; Cozzi, A.; Pageat, P.; Bienboire-Frosini, C. Neurohormonal Profiles of Assistance Dogs Compared to Pet Dogs: What Is the Impact of Different Lifestyles? *Animals* **2021**, *11*, 2594. [CrossRef] [PubMed]
12. Gazzano, A.; Ogi, A.; Macchioni, F.; Gatta, D.; Preziuso, G.; Baragli, P.; Curadi, M.C.; Giuliotti, L.; Sergi, V.; Casini, L. Blood serotonin concentrations in phobic dogs fed a dissociated carbohydrate-based diet: A pilot study. *Dog Behav.* **2019**, *5*, 9–17. [CrossRef]
13. Gazzano, A.; Casini, L.; Macchioni, F.; Mariti, C.; Baragli, P.; Curadi, M.C.; Giuliotti, L.; Riggio, G.; Ogi, A. L-Tryptophan supplementation increases serotonin blood levels in dogs fed a dissociated carbohydrate-based diet. *Dog Behav.* **2021**, *7*, 43–50. [CrossRef]
14. Piotti, P.; Uccheddu, S.; Alliani, M.; Mariti, C.; Nuti, V.; Ogi, A.; Pierantoni, L.; Gazzano, A. Management of specific fears and anxiety in the behavioral medicine of companion animals: Punctual use of psychoactive medications. *Dog Behav.* **2019**, *5*, 23–30. [CrossRef]
15. Ogi, A. A case of thunderstorm phobia in a Maremma sheepdog. *Dog Behav.* **2018**, *4*, 37–42. [CrossRef]
16. Gazzano, A.; Ogi, A.; Torracca, B.; Mariti, C.; Casini, L. Plasma Tryptophan/Large Neutral Amino Acids Ratio in Domestic Dogs Is Affected by a Single Meal with High Carbohydrates Level. *Animals* **2018**, *8*, 63. [CrossRef] [PubMed]
17. Mariti, C.; Falaschi, C.; Zilocchi, M.; Fatjó, J.; Sighieri, C.; Ogi, A.; Gazzano, A. Analysis of the intraspecific visual communication in the domestic dog (Canis familiaris): A pilot study on the case of calming signals. *J. Vet. Behav.* **2017**, *18*, 49–55. [CrossRef]
18. Mota-Rojas, D.; Marcet-Rius, M.; Ogi, A.; Hernández-Ávalos, I.; Mariti, C.; Martínez-Burnes, J.; Mora-Medina, P.; Casas, A.; Domínguez, A.; Reyes, B.; et al. Current advances in assessment of dog's emotions, facial expressions, and their use for clinical recognition of pain. *Animals* **2021**, *11*, 3334. [CrossRef] [PubMed]
19. Gazzano, A.; Giussani, S.; Gutiérrez, J.; Ogi, A.; Mariti, C. Attitude toward nonhuman animals and their welfare: Do behaviorists differ from other veterinarians? *J. Vet. Behav.* **2018**, *24*, 56–61. [CrossRef]
20. Casas-Alvarado, A.; Martínez-Burnes, J.; Mora-Medina, P.; Hernández-Avalos, I.; Domínguez-Oliva, A.; Lezama-García, K.; Gómez-Prado, J.; Mota-Rojas, D. Thermal and Circulatory Changes in Diverse Body Regions in Dogs and Cats Evaluated by Infrared Thermography. *Animals* **2022**, *12*, 789. [CrossRef] [PubMed]

Article

How Stressful Is Maternity? Study about Cortisol and Dehydroepiandrosterone-Sulfate Coat and Claws Concentrations in Female Dogs from Mating to 60 Days Post-Partum

Jasmine Fusi [1], Tanja Peric [2], Monica Probo [1,*], Alessio Cotticelli [2], Massimo Faustini [1] and Maria Cristina Veronesi [1]

[1] Department of Veterinary Medicine, Università degli Studi di Milano, via dell'Università, 6, 26900 Lodi, Italy; jasmine.fusi@unimi.it (J.F.); massimo.faustini@unimi.it (M.F.); maria.veronesi@unimi.it (M.C.V.)

[2] Department of Agricultural, Food, Environmental and Animal Sciences, University of Udine, Via Sondrio, 2/a, 33100 Udine, Italy; tanja.peric@uniud.it (T.P.); alessio.cotticelli@uniud.it (A.C.)

* Correspondence: monicaprobo@gmail.com

Simple Summary: Canine pregnancy and post-partum (total duration around 120 days) is a very intense period and could be a trigger for the activation of the Hypothalamic–Pituitary–Adrenal (HPA) axis in female dogs, given the higher allostatic load. To evaluate this activation, the concentrations of Cortisol (C) and Dehydroepiandrosterone-sulfate (DHEA-S), final products of the HPA axis, were detected in the coat and claws of 15 Dobermann Pinscher female dogs, with monthly collections from mating until the end of weaning. The C concentrations, both in the coat and claws, showed a significant trend of increase from mating until 60 days post-partum. In both matrices, the DHEA-S changes were not significant. Maternal parity and litter size did not play a significant influence on the concentrations of C and DHEA-S in both matrices. The results of the present study seem to depict maternity as a main activator of the HPA axis that, in turn, leads to the secretion of C. This is probably due to the increased allostatic load for the mothers, although it is not possible to discern the precise role of the multiple processes characterizing this period (uterine involution, lactation, nursing and grooming of the puppies).

Abstract: In dogs, the phase from mating to the end of weaning lasts about 120 days and encompasses many aspects that, interacting, contribute to increase the allostatic load. The coat and claws, useful for long-term change assessments, have the advantage of being collectable without invasiveness. In the present study, the Cortisol (C) and Dehydroepiandrosterone-sulfate (DHEA-S) concentration monthly changes in the coat and claws were studied in female dogs from mating to the end of weaning to assess Hypothalamic–Pituitary–Adrenal (HPA) axis activation during pregnancy and the post-partum period. The results from 15 Dobermann Pinscher female dogs showed a trend of increase of the coat C from mating to 60 days post-partum, with significant changes between mating and parturition-60 days post-partum ($p < 0.01$) and between the 30-day pregnancy diagnosis (PD) and 30–60 days post-partum ($p < 0.05$). The claws C trend showed significant increases between mating and 30–60 days post-partum ($p < 0.05$) and between the PD and 60 days post-partum ($p < 0.01$). DHEA-S in both matrices showed non-significant changes. The results suggest that maternity could play a pivotal role in the HPA axis activation, with a subsequent chronic secretion of C determining an increase in the allostatic load in the mothers. Neither maternal parity nor litter size played a significant role in the accumulation of C and DHEA-S in both matrices.

Keywords: dog; maternity; cortisol; dehydroepiandrosterone-sulfate; HPA axis

Citation: Fusi, J.; Peric, T.; Probo, M.; Cotticelli, A.; Faustini, M.; Veronesi, M.C. How Stressful Is Maternity? Study about Cortisol and Dehydroepiandrosterone-Sulfate Coat and Claws Concentrations in Female Dogs from Mating to 60 Days Post-Partum. *Animals* **2021**, *11*, 1632. https://doi.org/10.3390/ani11061632

Academic Editors: Angelo Gazzano and Asahi Ogi

Received: 27 April 2021
Accepted: 28 May 2021
Published: 31 May 2021

1. Introduction

In contrast to horse and cow newborns, newborn dog puppies are considered altricial and strictly dependent on the mother to survive until the end of weaning [1]. Canine pregnancies last about 63 ± 1 days after ovulation [2,3], and, at parturition, the phase of lactation, nursing and puppies grooming begins. In breeding facilities, the weaning of puppies starts at about 30 days after parturition, ending at 60 days post-partum. In Italy, puppies are usually sold to the new owners not earlier than 60 days of age and usually between 60 and 65 days of age. Therefore, the time elapsing between mating of the female dog and the end of weaning lasts about 120 days: altogether, this phase is complex and encompasses reproductive, metabolic, emotional and behavioral processes, often interacting among them, in turn increasing the allostatic load of the animal. This long-term period deserves scientific interest, as it represents a challenging and possibly very stressful time for the female dog [1,4]. Although many studies have focused on the specific phase of pregnancy, parturition and post-partum, to the authors' knowledge the whole period from mating until the end of weaning has only rarely been investigated in dogs [5]. Nonetheless, like what happens in horse husbandry, the selection of breeding female dogs is not based on the actual reproductive aptitude but, more often, on their show performances and genetic value. Due to the important role of maternal aptitude in puppies rearing, a more accurate selection of breeding female dogs could ameliorate the quality of dog breeding [6], contributing to limit the still too-high percentages of perinatal mortality [7] and conveying more ethical dog breeding. Still, today, it is not uncommon to observe disturbed behavior in periparturient female dogs, possible aggressiveness towards the newborns or even cannibalism, often accompanied by insufficient grooming or nursing of the puppies, thus impairing the successful outcome of litters. Among the diverse causes underlying these troubles, environmental factors, genetics, parity and number of the newborns in a litter were suggested [1,4,6]. Other than that, the role of the Hypothalamic–Pituitary–Adrenal axis (HPA axis) in shaping maternal behavior was presumed in sheep [8,9]. A study on adrenalectomized rats reported that, when the primiparae were injected with high doses of corticosterone, they showed higher levels of maternal care (notably, licking behavior) than the other primiparous rats who received lower doses of corticosterone [9,10]. Additionally, in humans, the HPA axis was reported to exert a role in mother–infant bonding [11]. In gorillas, a correlation between maternal behavior, stress and urinary Cortisol (C) concentrations was found; higher C concentrations were found in stressed mothers, whose infants were often less-cared-for [12]. In dogs, a similar association was presumed; mothers experienced peaks of salivary C after being temporarily separated from their litters, with higher peaks in mothers who displayed the most the typical maternal care behaviors [13]. It is thus evident that the role of stress is of utmost importance when evaluating data about the experience of maternity, both in humans and animals.

Lactation, especially when associated with nursing and grooming, is a challenging physiologic phase that requires a great expenditure of energy and could represent a source of stress in relation to the litter size [14] and to parental efforts. The increased metabolic load and stress are associated with activation of the HPA axis, leading to an increase of circulating C [14]. In a recent study on female cats, reference [14] reported that the highest concentrations of C in the blood were found 4 weeks after parturition, in association with the peak of lactation, suggesting a major effect of lactation and kitten nursing on HPA axis activation. Those authors concluded that measuring the circulating C concentrations could be helpful in deepening the understanding of the reaction of queens through the physiological stresses occurring during lactation.

Other than C, dehydroepiandrosterone (DHEA) and its sulfated form, dehydroepiandrosterone sulfate DHEA-S, are also final products of the activation of the HPA axis. Among multiple actions, DHEA is notably reported to be a neuroactive steroid with antidepressant actions, associated with some types of behaviors like sex recognition and aggressiveness [15–18]. In the literature, some results about the concentrations of circulating DHEA and DHEA-S during pregnancy exist in women [19], cows [20] and, more recently, in killer whales [21]. However, the results are conflicting: in women, a decline of circulating DHEA and DHEA-S during pregnancy was reported; in killer whales, a rise of DHEA-S from early and during mid and late pregnancy in comparison to pre-fertilization and post-partum was detected; lastly, a significant increase of circulating DHEA throughout pregnancy was detected in heifers and cows [19–21].

However, the HPA axis could not be considered as a "closed system", and its activity is modulated by factors such as oxytocin, able to exert an inhibitory action on the HPA axis. A recent study reported that the salivary oxytocin concentrations in female dogs were negatively correlated with excessive sniffing/poking behavior, considered as a possible sign of maternal distress associated with lactation [22]. Moreover, an allostatic theory of oxytocin was reported. According to this theory, the oxytocin system should be able to support the refinement of a physiologic setting to cope with adaptation and survival [23].

Despite most of the studies referring to concentrations of C and DHEA/DHEA-S in the blood or less invasive matrices such as urines, feces and saliva, all those matrices depict only the concentration of a given time point, and their use is therefore not suitable when long-lasting physiological processes have to be investigated because of the need to repeat multiple samplings. To overcome these issues, in the last years, new matrices, like the hair/coat and nails/claws, have been proposed for long-term hormonal studies in humans and animals [24–28] thanks to their characteristic of providing retrospective information about long-term hormone accumulation. Thus, the coat and claws could be suitable matrices for the study of hormonal changes occurring in the long-lasting phase between mating of the female dog to puppies weaning.

Therefore, the aim of the present study was to investigate the possible activation of the maternal HPA axis during the dynamic and challenging phase lasting from mating of the female dog to puppies weaning through the assessment of C and DHEA-S concentrations in the maternal coat and claws monthly collected.

2. Materials and Methods

2.1. Ethics

Although coat and claws sampling is a noninvasive procedure, this trial was carried out in accordance with EU Directive 2010/63/EU, and it was approved by the Ethical Committee of the University of Milan (OPBA_33_2021). Written informed consent was signed by the owners, giving permission to submit each female dog to elective C-section, to collect coat and claws samples and to allow the record of clinical data for research purposes.

2.2. Animals

Sixteen purebred Doberman Pinscher, 2nd to 3rd parity female dogs, aged 3–6 years old (mean ± SD: 4.1 ± 1.06), belonging to a single breeder were enrolled in the study. All the female dogs were found to be healthy by a clinical examination and by routine health tests performed by a veterinarian and submitted to the common vaccination and parasite prophylaxes, fed with commercial food and housed in a single kennel with indoor and outdoor spaces. The animals were always handled and managed by the same operator. All the female dogs showed a history of previous normal pregnancies, post-partum and lactation phases. In all of them, an elective cesarean section (ELCS) was performed at the previous whelping, because of the large litter sizes and the consequent risk for dystocia. Of the 16 mated female dogs enrolled in the present study, one was diagnosed as not pregnant at 30 days after ovulation and therefore removed from the study.

2.3. Estrus and Mating

From the onset of proestrus, all the female dogs were monitored through serial vaginal smears every 48–72 h and through the assay of plasma progesterone concentrations performed every 48 h from the beginning of the signs of cytological estrus. Based on these parameters, all the female dogs were submitted to a single mating, performed with a male of proven fertility, 48 h after the estimated ovulation, when progesterone plasma concentrations ranged between 4 and 10 ng/ml [29]. Each female dog was mated with a different healthy male of proven fertility. At the time of mating, the BCS of each female dog was assessed on a scale of 5.

2.4. Pregnancy, Caesarean Section, Newborn Puppies and Post-Partum Management

At 30 days after the estimated ovulation, pregnancy was checked by ultrasonographic examination, and the pregnant female dogs were submitted to measurement of the inner chorionic cavity (ICC) for the first calculation of the parturition date [30], while the nonpregnant female dogs were excluded from the study.

At 45 days after ovulation, a second ultrasonographic evaluation was performed to assess the normal course of pregnancy, the correct development and wellbeing of the fetuses and to measure the biparietal (BP) diameter for an additional calculation of the parturition date [30].

Given the reported higher risk for dystocia when the litter size counts more than 9 puppies [31], and because all the enrolled Dobermann Pinscher female dogs carried pregnancies with more than 9 fetuses, they were all scheduled for ELCS at term of pregnancy. The day for performing ELCS was scheduled on the basis of the concordance of several parameters, as previously reported [32–36]: date of ovulation, based on the measurement of plasma progesterone concentrations, ICC and BP. Moreover, during the last days of pregnancy, before the expected parturition date, the female dogs were checked daily to assess their clinical conditions and for ultrasonographic evaluations of the fetal wellbeing. Other than this, plasma progesterone concentrations were assessed to identify the pre-parturient decrease [37]. The ELCSs were performed only when plasma progesterone concentrations were <2 ng/ml.

For all the ELCSs, the same anesthetic and surgical protocols, aimed to minimize the possible negative effects on newborn viability and mothers' wellbeing, were performed, as *previously* reported [32,35,36].

Two expert neonatologists took care of the newborns as soon as they were extracted from the uterus, providing professional assistance. Within 5 minutes after birth, newborn viability was assessed by an Apgar score, and the puppies were classified as viable when the Apgar score was $\geq$7 [38]. The newborns were also evaluated for the absence of gross physical malformations or defects.

According to the viability classification, all the subjects were submitted to routine neonatal care or different degrees of neonatal assistance, as previously reported [38]. Birthweight, gender and litter size were also recorded.

Mothers and litters were discharged when female dogs were awake and showed normal behavior toward the puppies and after having verified the presence of normal mammary secretions.

A daily follow-up to update the general conditions was provided by the breeder from the day after parturition. Maternal and litter clinical data were daily recorded, together with the puppies' bodyweight until 60 days after parturition. In addition, at 1 and 2 weeks, and at 30 and 60 days after parturition, clinical examinations were scheduled. At 60 days after parturition, maternal BCS was reassessed.

2.5. Samples Collection

Coat samples were collected by using a razor (TN2300 Nomad, Rowenta® spa, Milan, Italy), allowing to shave an area of about 5 cm^2 from the dorsal surface of the right forearm until the level of the skin (coat samples). The tips of the claws were collected from all the digits of both forearms with a claw clipper (C135, Candure® Services LTD, Leicester, UK) and stored (claws samples). Coat and claws samples were placed in separate paper envelopes, labeled with a univocal code and stored at room temperature until the analysis.

The first sample collection was performed the day of mating. At the time of the pregnancy diagnosis, the second coat and claws sampling was performed, collecting only the regrown area.

At parturition, the third sample collection was performed. This collection did not always respect the 30 days of interval previously scheduled for the other samplings. Given that parturition occurs at about 60–62 days after ovulation, indeed, the interval of time between the second sampling and the one at parturition could have been 30 + a few days.

In all the cases, before anesthesia induction, the regrown coat and claws were collected and stored as reported above.

At 30 and 60 days post-partum, the regrown coat and claws were further collected and stored as reported above.

At every sampling time, after the collection, the razor and the claw clipper were disinfected with a 70% alcohol solution [39].

2.6. Hormone Analysis

Coat strands and claws were washed in 3-mL isopropanol to ensure the removal of any steroids on their surface. Coat and claws steroids were extracted with methanol and measured by radioimmunoassay (RIA). The concentrations of Cortisol and Dehydroepiandrosterone sulphate (DHEA-S) were measured using a solid-phase microtiter RIA. In brief, a 96-well microtiter plate (OptiPlate; PerkinElmer Life Sciences Inc., Zaventem, Belgium) was coated with goat anti-rabbit γ-globulin serum diluted 1:1.000 in 0.15-mM sodium acetate buffer (pH 9) and incubated overnight at 4 °C. The plate was then washed twice with RIA buffer (pH 7.5) and incubated overnight at 4 °C with 200 μL of the antibody serum diluted 1:20,000 for Cortisol and1:800 for DHEA-S. The cross-reactivities of the anti-Cortisol antibody with other steroids were as follows: Cortisol 100%, cortisone 4.3%, corticosterone 2.8%, 11-deoxycorticosterone 0.7%, 17-hydroxyprogesterone 0.6%, dexamethasone 0.1%, progesterone, 17-hydroxypregnenolone, DHEA-S, androsterone sulphate and pregnenolone < 0.01%. The cross-reactivities of the anti-DHEA-S antibody with other steroids were as follows: DHEA-S, 100%; androstenedione, 0.2%; DHEA, <0.01%; androsterone, <0.01% and testosterone, <0.01%. After washing the plate with RIA buffer, the standards (5–200 pg/well), the quality control extract, the test extracts and the tracer (hydrocortisone {Cortisol [1,2,6,7-3H (N)]-}, DHEA-S [1,2,6,7-3H (N)] were added, and the plate was incubated overnight at 4 °C. The bound hormone was separated from the free hormone by decanting and washing the wells in RIA buffer. After the addition of 200 μL of scintillation cocktail, the plate was counted on a β-counter (Top-Count; PerkinElmer Life Sciences Inc.).

The intra- and inter-assay coefficients of variation were 3.7 and 10.1% and 3.2 and 11.8% for Cortisol and DHEA-S, respectively. The sensitivities of the assays were 1.23 pg/well and 0.54 pg/well for Cortisol and DHEA-S, respectively.

2.7. Statistical Analysis

A Shapiro–Wilk test was used to verify the normal distribution of data, followed by an ANOVA and post-hoc test, to assess the possible effects of the sampling time, litter size, maternal age and parity on the C and DHEA-S concentrations in the coat and claw samples. Statistical significance was set for $p < 0.05$ (JASP®, ver. 9 for Windows platform). The Pearson correlation test was used to assess the possible correlations between the two matrices for each hormone.

3. Results

3.1. Clinical Findings

The 15 pregnant female dogs (BCS at mating, mean ± SD: 3.0 ± 0.00) showed normal courses of pregnancy and, because all of them carried more than nine fetuses, were submitted to the ELCS at 60–63 (mean ± SD: 61.2 ± 1.09) days after ovulation, providing a total of 15 litters with a total of 163 puppies (three stillborn). The litter sizes ranged between 10 and 13, with a mean ± SD of 10.9 ± 1.13. Seventy-five males and 85 females, mean ± SD birthweight: 427.7 ± 87.82 g, mean ± SD Apgar score: 8.9 ± 0.66, were enrolled. Consequently, all the 15 female dogs were followed for the whole period elapsing from mating until the end of weaning at 60 days post-partum, when the BCS was 2.5 ± 0.24 (mean ± SD).

3.2. Coat and Claws C and DHEA-S Concentrations

The concentrations (mean ± SD) of C and DHEA-S in the coat and claws from mating to 60 days post-partum in the 15 female dogs enrolled in the study are reported in Figures 1 and 2, respectively.

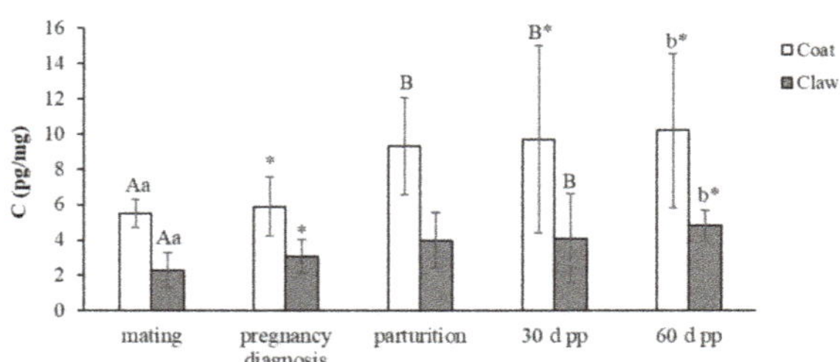

Figure 1. Concentrations (mean ± SD) of C in the coats and claws from mating to 60 days post-partum in the 15 female dogs. ^{A,B} $p < 0.05$; ^{a,b} $p < 0.01$ denote within-row significance. * $p < 0.05$ denotes within-row significance.

The concentrations of C in the coats showed a trend of increase from mating to 60 days post-partum, with significant changes between mating and parturition-60 days post-partum and between the pregnancy diagnosis and 30–60 days post-partum. In the claws, the trend of C was very similar to the one observed in the coats, with significant increases between mating and 30–60 days post-partum and between pregnancy diagnosis and 60 days post-partum.

The Pearson correlation test showed a significant positive correlation (r = 0.277; $p < 0.05$) between the coat and claw C concentrations.

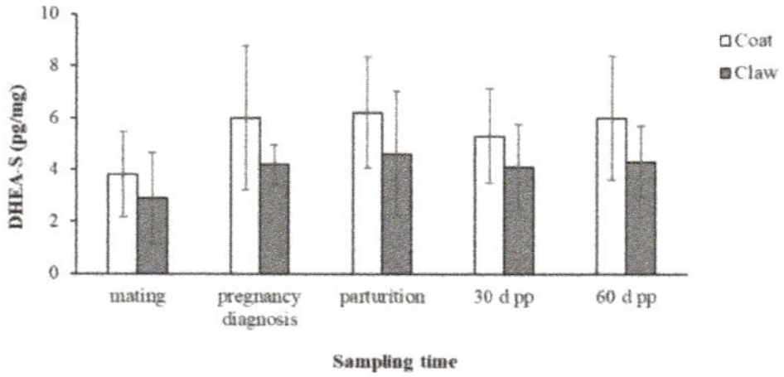

Figure 2. Concentrations (mean ± SD) of DHEA-S in the coats and claws from mating to 60 days post-partum in the 15 female dogs.

The concentrations of DHEA-S in both the coats and claws showed the same trend of increase between mating and all the subsequent sampling times but without significant changes.

The statistical analysis showed the absence of significant effects of the maternal age and litter size on the C and DHEA-S concentrations in the coats and claws, and no significant correlations of DHEA-S between the two matrices was found.

4. Discussion

To the authors' knowledge, this was the first study reporting the concentrations of C and DHEA-S- in the coat and claws of female dogs from mating to 60 days post-partum, providing new information about long-term hormonal changes during gestation and the post-partum period in dogs, a topic that needs to be further investigated also in humans [40].

Some studies previously demonstrated the usefulness of a single coat sample for the study of C [25] and DHEA-S [41] concentrations in dogs. However, the present study offered the first evidence that, using the shave–re-shave method [27,42–44] after the first sampling, it is possible to collect only the regrown matrices, providing an optimal tool to investigate the long-term hormonal variations in dogs, as previously reported for other species [45–47]. However, in valuable dogs, the owners often refuse to shave the hair, even in a small area, limiting the use of the coat as a matrix of study in many instances. A couple of recent studies [26,28] showed the usefulness of the claws as an alternative to the coat for the same long-term studies in newborn puppies. For these reasons, in the present study, both the coat and the claws were collected with the shave—re-shave and clip—re-clip method for the longitudinal analysis of C and DHEA-S long-term changes.

The 30-day collection interval was designed to address several targets. Firstly, it reduced, at the minimum, the number of samplings to restrict the possible disturbances for the pregnant/lactating female dogs. Secondly, this timing fits with the most important milestones of the reproductive process management in female dogs. Thirdly, it allowed the regrowth of a suitable amount of sample for the analysis. Fourthly, it was reasonable for identifying and studying long-term hormonal changes. About the area of the body chosen for the coat collection, the right forearm was used because it is shaved for blood collection and other clinical purposes.

About the enrollment criteria of the female dogs, in the present study, only one breed was selected to avoid possible genetic differences in HPA axis activation [48]. However, even within the same breed, some individual peculiarities in HPA axis activation could be presumed. Moreover, to avoid possible differences in the managerial, nutritional and environmental factors that could influence HPA axis activation, all the female dogs belonged to the same breeder and were managed by the same operator. Other than this, selecting a single breed and the consequent homogeneous litter size, together with similar parity, allowed an additional reduction of the possible variables in HPA axis activation. The choice of enrolling only female dogs of second and third parity was based on the hypothesized possible influence of the maternal experience in HPA axis activation, as reported for maternal behavior [6]. The exclusive inclusion of female dogs scheduled for ELCS, in addition, concurred to reduce the other possible variables related to the effects of different types of delivery on HPA axis activation, as previously observed [49].

About C, the trend of increase from mating to 60 days post-partum observed in both the coat and claws suggested a gradual activation of the maternal HPA axis during the delicate phase of pregnancy and post-partum in female dogs. The finding of the highest concentration of C at 60 days post-partum is intriguing and could be related to the complex interaction of multiple factors, such as uterine involution, lactation and, also, nursing and grooming of the puppies, therefore addressing not only metabolic and physiologic phenomena but, also, emotional and behavioral changes for female dogs. All these factors, in turn, play a role in the increase of the allostatic load for female dogs. Although it is not possible to discern the single contribution of each one of these processes on the activation of the HPA axis, it is possible to suppose that lactation and care of the puppies could promote a continuous stimulation on the HPA axis, leading to a chronic secretion of C, as evidenced by its long-term, retrospective accumulation in the coat and claws. In the present study, all the female dogs displayed normal maternal behavior in the present and previous pregnancies, so that the observed HPA axis activation could be supposed to reflect a "normal" process related to maternity in dogs. However, it is not possible to exclude that the energy expenditure needed for all of the processes above seen could stimulate C secretion in order to mobilize the depots, as supposed in reference [14], given the higher concentrations reported at the peak of lactation.

The statistical analysis did not show an effect of maternal parity on C coat and claw concentrations. This could be due to very limited differences in the parity of the enrolled female dogs. However, in the bottlenose dolphin, the profiles of C concentrations during pregnancy were not affected by age or parity [50], and the same results were reported for killer whales [21]. In a recent study, the authors of reference [14] reported that the maternal experience did not represent an influencing factor on C concentrations in the blood of queens at 4 weeks after parturition, while it was influenced by the size of the litter (higher in queens with up to three kittens than in queens with more than three kittens). In the present study, the litter size did not influence the C concentrations in the coats and claws. It should be noted that, in the enrolled female dogs, no extreme litter sizes were observed, and the litter sizes were rather homogeneous, with the number of puppies per litter ranging between 10 and 13. Moreover, about the possible effect played by the puppies on the maternal HPA axis activation, in the present study, excluding the three stillborn, the puppies were all healthy and viable at birth and for all the subsequent periods of observation, showing normal growth and development in the first 60 days of age.

Therefore, taking all these considerations together, the results of the present study should be considered representative of the "normal" conditions of both mothers and litters. Hence, it could be interesting to verify the possible differences of long-term maternal HPA axis activation according to breed, first parity, extreme litters, pathological situations of mothers and/or puppies, etc.

Beside the C concentrations in the coats and claws, it was unforeseen to observe no significant changes in the concentrations of DHEA-S in both matrices. Even if conflicting results are reported in women, cows and killer whales, the role of DHEA as a neuroactive steroid is known to be involved in behavioral changes [15–18], in stress responses [15–17] and in the stimulation of lobuloalveolar mammary gland development in rats [51]. Like C, the concentrations of DHEA-S in the coat and claws were not influenced by the maternal parity or litter size.

In a study by Fusi and colleagues [26], both the C and DHEA(S) concentrations measured in the claws collected from puppies since birth to 60 days of age were found to be higher in claws collected at birth than in the subsequent sampling times, with a trend of decreasing. Therefore, because no differences related to DHEA-S claw concentrations were found in the present study in female dogs, it could be supposed that DHEA-S is mainly synthesized by the fetus/newborn. This hypothesis could also be reinforced by the extremely different concentrations found in the present study in the claws of female dogs compared to the study in reference [26] in the claws of puppies when the same time from parturition to 60 days post-partum was concerned. In fact, in the study from reference [26], the DHEA(S) claw concentrations were about 50-fold higher in puppies at parturition, about 20-fold higher at 30 days post-partum and about 15-fold higher at 60 days post-partum, as compared to the female dogs investigated in the present study.

Additionally, about C, higher concentrations were found in the puppies of the same study when compared to the concentrations found in the female dogs of the present one, especially at parturition (about six-fold).

These comparisons seem to suggest that, in dogs, the fetal HPA axis activation plays an important role, especially around the time of parturition.

The results of the present study offer new insights in the role of maternity on the allostatic load of female dogs. This topic is of maximum importance, given the influence of the allostatic load on the health of an organism, as reported in humans [52], and suggests that, although it is a physiologic event, the periods of pregnancy and, especially, post-partum should be considered as possible stressors for female dogs.

5. Conclusions

In conclusion, the higher concentration of C in the coat and claws found at 60 days post-partum than those found at mating seems to suggest that a complex interplay among the physiologic, emotional and behavioral events in the post-partum period occurs in female dogs. Maternity seems to play a pivotal role in the HPA axis activation in female dogs, with a subsequent chronic secretion of C and long-term accumulation in the coat and claws. No significant differences were found in the DHEA-S concentrations in the different sampling times, and neither maternal parity nor litter size played a significant role in the accumulation of C and DHEA-S in both matrices.

Author Contributions: Conceptualization, M.C.V. and J.F.; methodology, T.P. and M.P.; software, M.F.; validation, A.C. and T.P.; formal analysis, M.F.; investigation, J.F. and M.P.; data curation, M.C.V. and J.F.; writing—original draft preparation, M.P. and J.F.; writing—review and editing, M.C.V. and J.F. and supervision, M.C.V. All authors have read and agreed to the published version of the manuscript.

Funding: This research was funded by the Università degli Studi di Milano (UNIMI), grant number LINEA2_CVERO_2019_AA.

Institutional Review Board Statement: This study was conducted according to the guidelines of the Declaration of Helsinki and approved by the Institutional Review Board of the Università degli Studi di Milano-Organismo Preposto al Benessere degli Animali (OPBA) (protocol code OPBA_33_2021).

Informed Consent Statement: Not applicable.

Data Availability Statement: The data are available upon request from the authors.

Acknowledgments: The authors are grateful to the breeder Chiara Campazzo for her support.

Conflicts of Interest: The authors declare no conflict of interest.

References

1. Lezama-García, K.; Mariti, C.; Mota-Rojas, D.; Martínez-Burnes, J.; Barrios-García, H.; Gazzano, A. Maternal behaviour in domestic dogs. *Int. J. Vet. Sci. Med.* **2019**, *7*, 20–30. [CrossRef] [PubMed]
2. Concannon, P.W.; Whaley, S.; Lein, D.; Wissler, R. Canine gestation length: Variation related to time of mating and fertile life of sperm. *Am. J. Vet. Res.* **1983**, *44*, 1819–1821. [PubMed]
3. Kutzler, M.A.; Mohammed, H.O.; Lamb, S.V.; Meyers-Wallen, V.N. Accuracy of canine parturition date prediction from the initial rise in preovulatory progesterone concentration. *Theriogenology* **2003**, *60*, 1187–1196. [CrossRef]
4. Santos, N.R.; Beck, A.; Blondel, T.; Maenhoudt, C.; Fontbonne, A. Influence of dog-appeasing pheromone on canine maternal behaviour during the peripartum and neonatal periods. *Vet. Rec.* **2020**, *186*, 449. [CrossRef]
5. Cardinali, L.; Troisi, A.; Verstegen, J.P.; Menchetti, L.; Elad Ngonput, A.; Boiti, C.; Canello, S.; Zelli, R.; Polisca, A. Serum concentration dynamic of energy homeostasis hormones, leptin, insulin, thyroid hormones, and cortisol throughout canine pregnancy and lactation. *Theriogenology* **2017**, *97*, 154–158. [CrossRef]
6. Santos, N.R.; Beck, A.; Fontbonne, A. A review of maternal behaviour in dogs and potential areas for further research. *J. Small Anim. Pract.* **2020**, *61*, 85–92. [CrossRef]
7. Tønnessen, R.; Borge, K.S.; Nødtvedt, A.; Indrebø, A. Canine perinatal mortality: A cohort study of 224 breeds. *Theriogenology* **2012**, *77*, 1788–1801. [CrossRef]
8. Keverne, E.B.; Kendrick, K.M. Morphine and corticotrophin-releasing factor potentiate maternal acceptance in multiparous ewes after vaginocervical stimulation. *Brain Res.* **1991**, *540*, 55–62. [CrossRef]
9. Lévy, F. Neuroendocrine control of maternal behavior in non-human and human mammals. *Ann. Endocrinol.* **2016**, *77*, 114–125. [CrossRef]
10. Rees, S.L.; Panesar, S.; Steiner, M.; Fleming, A.S. The effects of adrenalectomy and corticosterone replacement on maternal behavior in the postpartum rat. *Horm. Behav.* **2004**, *46*, 411–419. [CrossRef]
11. Mastorakos, G.; Ilias, I. Maternal hypothalamic-pituitary-adrenal axis in pregnancy and the postpartum period. Postpartum-related disorders. *Ann. N. Y. Acad. Sci.* **2000**, *900*, 95–106. [CrossRef]
12. Bahr, N.I.; Pryce, C.R.; Döbeli, M.; Martin, R.D. Evidence from urinary cortisol that maternal behavior is related to stress in gorillas. *Physiol. Behav.* **1998**, *64*, 429–437. [CrossRef]
13. Bray, E.E.; Sammel, M.D.; Cheney, D.L.; Serpell, J.A.; Seyfarth, R.M. Characterizing Early Maternal Style in a Population of Guide Dogs. *Front. Psychol.* **2017**, *8*, 175. [CrossRef]
14. Alekseeva, G.S.; Loshchagina, J.A.; Erofeeva, M.N.; Naidenko, S.V. Stressed by Maternity: Changes of Cortisol Level in Lactating Domestic Cats. *Animals* **2020**, *10*, 903. [CrossRef]
15. Baulieu, E.E. Neurosteroids: A novel function of the brain. *Psychoneuroendocrinology* **1998**, *23*, 963–987. [CrossRef]
16. Zinder, O.; Dar, D.E. Neuroactive steroids: Their mechanism of action and their function in the stress response. *Acta Physiol. Scand.* **1999**, *167*, 181–188. [CrossRef]

17. Dubrovsky, B. The specificity of stress responses to different nocuous stimuli: Neurosteroids and depression. *Brain Res. Bull.* **2000**, *51*, 443–455. [CrossRef]
18. Maurice, T.; Urani, A.; Phan, V.L.; Romieu, P. The interaction between neuroactive steroids and the sigma1 receptor function: Behavioral consequences and therapeutic opportunities. *Brain Res. Rev.* **2001**, *37*, 116–132. [CrossRef]
19. Bird, C.E.; Cook, S.J.; Dressel, D.E.; Clark, A.F. Plasma C19-delta 5-steroid levels during normal human pregnancy. *Clin. Biochem.* **1980**, *13*, 126–128. [CrossRef]
20. Gabai, G.; Marinelli, L.; Simontacchi, C.; Bono, G.G. The increase in plasma C19Delta5 steroids in subcutaneous abdominal and jugular veins of dairy cattle during pregnancy is unrelated to estrogenic activity. *Steroids* **2004**, *69*, 121–127. [CrossRef]
21. Robeck, T.R.; Steinman, K.J.; O'Brien, J.K. Characterization and longitudinal monitoring of serum androgens and glucocorticoids during normal pregnancy in the killer whale (Orcinus orca). *Gen. Comp. Endocrinol.* **2017**, *247*, 116–129. [CrossRef]
22. Ogi, A.; Mariti, C.; Pirrone, F.; Baragli, P.; Gazzano, A. The Influence of Oxytocin on Maternal Care in Lactating Dogs. *Animals* **2021**, *11*, 1130. [CrossRef]
23. Quintana, D.S.; Guastella, A.J. An Allostatic Theory of Oxytocin. *Trends Cogn. Sci.* **2020**, *24*, 515–528. [CrossRef]
24. Tegethoff, M.; Raul, J.S.; Jamey, C.; Khelil, M.B.; Ludes, B.; Meinlschmidt, G. Dehydroepiandrosterone in nails of infants: A potential biomarker of intrauterine responses to maternal stress. *Biol. Psychol.* **2011**, *87*, 414–420. [CrossRef]
25. Veronesi, M.C.; Comin, A.; Meloni, T.; Faustini, M.; Rota, A.; Prandi, A. Coat and claws as new matrices for noninvasive long-term cortisol assessment in dogs from birth up to 30 days of age. *Theriogenology* **2015**, *84*, 791–796. [CrossRef]
26. Fusi, J.; Comin, A.; Faustini, M.; Prandi, A.; Veronesi, M.C. The usefulness of claws collected without invasiveness for cortisol and dehydroepiandrosterone (sulfate) monitoring in healthy newborn puppies after birth. *Theriogenology* **2018**, *122*, 137–143.
27. Heimbürge, S.; Kanitz, E.; Otten, W. The use of hair cortisol for the assessment of stress in animals. *Gen. Comp. Endocrinol.* **2019**, *270*, 10–17. [CrossRef]
28. Fusi, J.; Comin, A.; Faustini, M.; Prandi, A.; Veronesi, M.C. Perinatal concentrations of 17β-estradiol and testosterone in the toe claws of female and male dogs from birth until 60 days of age. *Anim. Reprod. Sci.* **2020**, *214*, 106313. [CrossRef]
29. Levy, X.; Fontbonne, A. Determining the optimal time of mating in bitches: Particularities. *Rev. Bras. Reprod. Anim.* **2007**, *31*, 128–134.
30. Alonge, S.; Beccaglia, M.; Melandri, M.; Luvoni, G.C. Prediction of whelping date in large and giant canine breeds by ultrasonography foetal biometry. *J. Small Anim. Pract.* **2016**, *57*, 479–483. [CrossRef]
31. Cornelius, A.J.; Moxon, R.; Russenberger, J.; Havlena, B.; Cheong, S.H. Identifying risk factors for canine dystocia and stillbirths. *Theriogenology* **2019**, *128*, 201–206. [CrossRef] [PubMed]
32. Meloni, T.; Comin, A.; Rota, A.; Peric, T.; Contri, A.; Veronesi, M.C. IGF-I and NEFA concentrations in fetal fluids of term pregnancy dogs. *Theriogenology* **2014**, *81*, 1307–1311. [CrossRef] [PubMed]
33. Dall'Ara, P.; Meloni, T.; Rota, A.; Servida, F.; Filipe, J.; Veronesi, M.C. Immunoglobulins G and lysozyme concentrations in canine fetal fluids at term of pregnancy. *Theriogenology* **2015**, *83*, 766–771. [CrossRef] [PubMed]
34. Bolis, B.; Prandi, A.; Rota, A.; Faustini, M.; Veronesi, M.C. Cortisol fetal fluid concentrations in term pregnancy of small-sized purebred dogs and its preliminary relation to first 24 hours survival of newborns. *Theriogenology* **2017**, *88*, 264–269. [CrossRef] [PubMed]
35. Veronesi, M.C.; Bolis, B.; Faustini, M.; Rota, A.; Mollo, A. Biochemical composition of fetal fluids in at term, normal developed, healthy, viable dogs and preliminary data from pathologic littermates. *Theriogenology* **2018**, *108*, 277–283. [CrossRef] [PubMed]
36. Fusi, J.; Faustini, M.; Bolis, B.; Veronesi, M.C. Apgar score or birthweight in Chihuahua dogs born by elective Caesarean section: Which is the best predictor of the survival at 24 h after birth? *Acta Vet. Scand.* **2020**, *62*, 39. [CrossRef] [PubMed]
37. Concannon, P.W. Canine pregnancy: Predicting parturition and timing events of gestation. In *Recent Advances in Small Animal Reproduction*; Concannon, P.W., England, G., Verstegen, J., Linde-Forsberg, C., Eds.; International Veterinary Information Service: Ithaca, NY, USA, 2000; A1202.0500; Available online: http://www.ivis.org (accessed on 6 April 2017).
38. Veronesi, M.C.; Panzani, S.; Faustini, M.; Rota, A. An Apgar scoring system for routine assessment of newborn puppy viability and short-term survival prognosis. *Theriogenology* **2009**, *72*, 401–407. [CrossRef]
39. Baier, F.; Grandin, T.; Engle, T.; Edwards-Callaway, L. Evaluation of Hair Characteristics and Animal Age on the Impact of Hair Cortisol Concentration in Feedlot Steers. *Front. Vet. Sci.* **2019**, *6*, 323. [CrossRef]
40. Kuijper, E.A.; Ket, J.C.; Caanen, M.R.; Lambalk, C.B. Reproductive hormone concentrations in pregnancy and neonates: A systematic review. *Reprod. Biomed. Online* **2013**, *27*, 33–63. [CrossRef]
41. Bolis, B.; Peric, T.; Veronesi, M.C.; Faustini, M.; Rota, A.; Montillo, M. Hair dehydroepiandrosterone (DHEA) concentrations in newborn dogs. In Proceedings of the XVIII EVSSAR Congress, Hannover, Germany, 11–12 September 2015; p. 124.
42. Davenport, M.D.; Tiefenbacher, S.; Lutz, C.K.; Novak, M.A.; Meyer, J.S. Analysis of endogenous cortisol concentrations in the hair of rhesus macaques. *Gen. Comp. Endocrinol.* **2006**, *147*, 255–261. [CrossRef]
43. Meyer, J.S.; Novak, M.A. Minireview: Hair cortisol: A novel biomarker of hypothalamic-pituitary- adrenocortical activity. *Endocrinology* **2012**, *153*, 4120–4127. [CrossRef]
44. Mesarčová, L.; Kottferová, J.; Skurková, L.; Lešková, L.; Kmecová, N. Analysis of cortisol in dog hair—A potential biomarker of chronic stress: A review. *Vet. Med.* **2017**, *62*, 363–376. [CrossRef]
45. Comin, A.; Peric, T.; Montillo, M.; Faustini, M.; Zufferli, V.; Cappa, A.; Cornacchia, G.; Prandi, A. Hair Cortisol Levels to Monitor Hypothalamic-Pituitary-Adrenal Axis Activity in Healthy Dairy Cows. *J. Anim. Vet. Adv.* **2012**, *11*, 3623–3626.

46. Comin, A.; Veronesi, M.C.; Montillo, M.; Faustini, M.; Valentini, S.; Cairoli, F.; Prandi, A. Hair cortisol level as a retrospective marker of hypothalamic-pituitary-adrenal axis activity in horse foals. *Vet. J.* **2012**, *194*, 131–132. [CrossRef]
47. Comin, A.; Zufferli, V.; Peric, T.; Canavese, F.; Barbetta, D.; Prandi, A. Hair cortisol levels determined at different body sites in the New Zealand White rabbit. *World Rabbit Sci.* **2012**, *20*, 149–154. [CrossRef]
48. Potischman, N.; Troisi, R.; Thadhani, R.; Hoover, R.N.; Dodd, K.; Davis, W.W.; Sluss, P.M.; Hsieh, C.C.; Ballard-Barbash, R. Pregnancy hormone concentrations across ethnic groups: Implications for later cancer risk. *Cancer Epidemiol. Biomark. Prev.* **2005**, *14*, 1514–1520. [CrossRef]
49. Fusi, J.; Carluccio, A.; Peric, T.; Faustini, M.; Prandi, A.; Veronesi, M.C. Effect of Delivery by Emergency or Elective Cesarean Section on Nitric Oxide Metabolites and Cortisol Amniotic Concentrations in at Term Normal Newborn Dogs: Preliminary Results. *Animals* **2021**, *11*, 713. [CrossRef]
50. Steinman, K.J.; Robeck, T.R.; O'Brien, J.K. Characterization of estrogens, testosterone, and cortisol in normal bottlenose dolphin (Tursiops truncatus) pregnancy. *Gen. Comp. Endocrinol.* **2016**, *226*, 102–112. [CrossRef]
51. Sourla, A.; Martel, C.; Labrie, C.; Labrie, F. Almost exclusive androgenic action of dehydroepiandrosterone in the rat mammary gland. *Endocrinology* **1998**, *139*, 753–764. [CrossRef]
52. Guidi, J.; Lucente, M.; Sonino, N.; Fava, G.A. Allostatic Load and Its Impact on Health: A Systematic Review. *Psychother. Psychosom.* **2021**, *90*, 11–27. [CrossRef]

 animals

Article

Blood Biomarker Profile Alterations in Newborn Canines: Effect of the Mother's Weight

Brenda Reyes-Sotelo [1], Daniel Mota-Rojas [2,*], Patricia Mora-Medina [3], Asahi Ogi [4], Chiara Mariti [4], Adriana Olmos-Hernández [5], Julio Martínez-Burnes [6], Ismael Hernández-Ávalos [3], Jose Sánchez-Millán [3] and Angelo Gazzano [4]

1 Science Program "Maestria en Ciencias Agropecuarias", Xochimilco Campus, Universidad Autónoma Metropolitana, Mexico City 04960, Mexico; breyess_20@yahoo.com.mx
2 Neurophysiology, Behavior and Animal Welfare Assesment, DPAA, Xochimilco Campus, Universidad Autónoma Metropolitana, Mexico City 04960, Mexico
3 Facultad de Estudios Superiores Cuautitlán, Universidad Nacional Autónoma de Mexico (UNAM), Cuautitlán Izcalli 54714, Mexico; mormed2001@yahoo.com.mx (P.M.-M.); mvziha@hotmail.com (I.H.-Á.); jlsan@unam.mx (J.S.-M.)
4 Department of Veterinary Sciences, University of Pisa, 56124 Pisa, Italy; asahi.ogi@vet.unipi.it (A.O.); chiara.mariti@unipi.it (C.M.); angelo.gazzano@unipi.it (A.G.)
5 Division of Biotechnology-Bioterio and Experimental Surgery, Instituto Nacional de Rehabilitación Luis Guillermo Ibarra Ibarra (INR-LGII), Mexico City 14389, Mexico; adrianaolmos05@gmail.com
6 Animal Health Group, Facultad de Medicina Veterinaria y Zootecnia, Universidad Autónoma de Tamaulipas, Victoria City 87000, Mexico; jmburnes@docentes.uat.edu.mx
* Correspondence: dmota100@yahoo.com.mx

Citation: Reyes-Sotelo, B.; Mota-Rojas, D.; Mora-Medina, P.; Ogi, A.; Mariti, C.; Olmos-Hernández, A.; Martínez-Burnes, J.; Hernández-Ávalos, I.; Sánchez-Millán, J.; Gazzano, A. Blood Biomarker Profile Alterations in Newborn Canines: Effect of the Mother's Weight. *Animals* **2021**, *11*, 2307. https://doi.org/10.3390/ani11082307

Academic Editor: Michael Lindinger

Received: 29 June 2021
Accepted: 28 July 2021
Published: 5 August 2021

Simple Summary: Morphological variability in canines is associated with the mother's size and weight, which likely affects the birth weight of the puppies and their metabolic status. Identifying physio-metabolic alterations in the blood from the umbilical vein to evaluate the concentration of gases, glucose, lactate, calcium, hematocrit levels, and blood pH of newborn puppies will make it possible to determine the risk of complications due to intrauterine asphyxia. The objective of this study is to evaluate the effect of the mother's weight on the weight of liveborn and stillborn puppies during spontaneous births and the neonates' blood physiological alterations during the first minute of life. The above allowed us to identify the physio-metabolic maladjustments that newborn puppies suffer from and to determine the risk of asphyxia according to the weight category of the mothers. Results suggest that if the weight of the bitch is >16.1 kg in eutocic births, there is a higher risk of intrapartum physiological alterations and death. The results of this study allowed us to identify that the weight of dams before birth determines the weight of the puppies at birth, though there is a wide range in birth weights due to the ample morphological variability characteristics of this species.

Abstract: This study aims to determine the effect of the weight of bitches on liveborn and stillbirth puppies from eutocic births, and physiological blood alterations during the first minute postpartum. A total of 52 female dogs were evaluated and distributed in four categories: C1 (4.0–8.0 kg, n = 19), C2 (8.1–16.0 kg, n = 16), C3 (16.1–32.0 kg, n = 11), and C4 (32.1–35.8 kg, n = 6). The dams produced 225 liveborn puppies and 47 were classified as stillbirth type II. Blood samples were taken from the umbilical vein to evaluate the concentration of gases, glucose, lactate, calcium, hematocrit levels, and blood pH. The liveborn puppies in C2, C3, and C4 had more evident physiological alterations (hypercapnia, acidosis) than those in C1 ($p < 0.05$). These signs indicate a process of transitory asphyxiation. The stillborn pups in all four categories had higher weights than their liveborn littermates. C3 and C4 had the highest mean weights (419.86 and 433.79 g, respectively) and mortality rates (C3 = 20.58%, C4 = 24.58%). Results suggest that if the weight of the bitch is >16.1 kg in eutocic births, there is a higher risk of intrapartum physiological alterations and death. The results of this study allowed us to identify that the weight of dams before birth determines the weight of the puppies at birth.

Keywords: animal perinatology; asphyxia; physiological blood profile; puppy welfare; stillbirth

1. Introduction

Mortality in dogs during the neonatal period has been estimated to reach 40% [1]. Deaths may occur in the uterus, during expulsion, immediately postpartum, or during the first weeks of life [2–4], but the highest number of stillbirths occurs during birth [5] and the first 7 days of life [6]. Approximately 60% of these deaths are associated with intrapartum asphyxiation [7] caused by dystocic deliveries [5,6,8]. Asphyxia during the birthing process also negatively impacts the newborns' adaptation to extrauterine life [9] by limiting their viability and vitality [6,10–13]. A high neurologic morbidity increases the risk of neonatal mortality [14]. The birthing process is the most critical phase for newborns [15] because the transition from fetus to neonate involves physiological, biochemical, and anatomical changes accompanied by flows of hormones that trigger the respiratory function, vascular changes, and the activation of energy metabolism [16,17]; additionally, the maternal behavior is critical for the parturition to take place in favorable conditions for the newborn puppy [18–21]. Studies of dogs have reported that a certain level of transitory asphyxiation occurs during delivery. Though this is normal, it produces hypercapnia and transitory acidosis in puppies [22,23]. If these conditions persist, they will alter gas exchange [24], delay the onset of respiration, and generate metabolic acidosis in newborns [25]. The challenges of the birthing process, together with these risk factors can determine the proportion of the liveborn (LP) vs. stillbirth (SB) puppies and the viability of the former [26–28]. Morphological variability in canines is associated with the mother's size and weight [23], for these likely affect the birth weight of the puppies [26,28–30] and their metabolic status. In both veterinary and human perinatology, analyzing blood gases and metabolites has emerged as an important tool for evaluating newborns [13,31], but reports on dogs are scarce. Studying physiological indicators provides crucial information and allows researchers to estimate variations in oxygenation levels, metabolic profiles, and the acid–base balance [32] that help determine the level of fetal hypoxia suffered during birth. Gasometry allows the monitoring of the respiratory function by measuring the concentration of certain gases (pO_2, O_2 saturation (SaO_2), pCO_2) and blood pH [11,12,33–36] and the evaluation of the acid-base balance—to estimate the newborns' metabolic status [13,33,37,38]. Variations in metabolite levels, including lactate, play an important role in metabolic acidosis [39,40] associated with hypoxic events [1,41], high blood glucose levels [36], and a general compensatory metabolism marked by excess base and bicarbonate in the blood [25]. Identifying physio-metabolic alterations in the blood of newborn puppies will make it possible to determine the risk of complications due to intrauterine asphyxia. However, evidence on hypoxia in canines, its effects, and its relations to the mother's weight as a risk factor is scant or has not been fully evaluated. Thus, the objective of this study was to evaluate the effect of the mother's weight on the weight of the LP and SB puppies during spontaneous births and the neonates' blood physio-metabolic alterations during the first minute of life. The above allowed us to identify the physiological maladjustments that newborn puppies suffer and determine the risk of asphyxia according to the weight category of the mothers.

2. Materials and Methods

2.1. Infrastructure

A network of 10 veterinary clinics in Mexico City was organized to recruit pregnant dogs. Prenatal control was performed from day 25 of pregnancy to 24 h postpartum.

2.2. Study Population

A total of 52 young multiparous bitches (2–4 births) were recruited. The inclusion criteria were: (i) clinically healthy dogs; (ii) valid vaccination/deworming record; (iii) fed a commercial formula; (iv) no history of reproductive problems; (v) radiographic and ultrasonographic evaluations to show they were apt for natural births. The exclusion criteria were: (i) primiparous females; (ii) bitches with a history of dystocia or pyometra; (iii) previous type I SB puppies; (iv) malformed fetuses; (v) use of birth inducers or accelerators; (vi) bitches with body condition 8 or 9 (obese) as per the WSAVA scale [42]; (vii) extremely aggressive behavior. Brachycephalic and large breeds were excluded due to their reported high incidence of dystocia [3]. The 52 pregnant females were classified in 4 categories according to their weight recorded before labor (day 60 $\pm$ 2), as follows: C1 (4.0–8.0 kg, n = 19); C2 (8.1–16.0 kg, n = 16); C3 (16.1–32.0 kg, n = 11); C4 (32.1–35.8 kg, n = 6). The body weight ranges respected the general guidelines for breed size established by the Federation Cynologique Internationale (FCI) [43].

2.3. Clinical History

The clinical history of the dogs was compiled, including age, weight, alimentation, preventive medicine status, and a description of the environment where they lived. All information was recorded in the Q.vet® Ed. Professional 2016 database for veterinary clinics.

2.4. Diagnoses of Pregnancy

Diagnoses were confirmed between days 24 and 28 post-service for each dam. Fetal structures and cardiac activity were detected in the gestational sacs using a LOGIQ 400 MD ultrasound machine (General Electric, Yokohama, Japan) equipped with a 3.5 MHz convex transducer to establish probable due dates. Monitoring of fetal maturation and vitality was performed on days 40–50 of gestation. The fetal structure was defined completely to permit the early identification of pyometra cases, type I SB puppies, and malformations. X-rays of the dams' abdomens were taken on day 45 of pregnancy once bone calcification of the fetuses was achieved to discard early stages of maternal–fetal dystocia and evidence of cephalopelvic disproportion, conditions that would make caesarean sections necessary [29]; thus, excluding the dam from the study. On day 60 of pregnancy, the females were checked by ultrasound to corroborate cardiac rhythms and fetal biparietal diameters. Monitoring of births was performed with a model S80Vet Sino-Hero® vital sign monitor to evaluate the mothers' physiological parameters. Clinical signs observed in the peripartum interval included anorexia, anxiety, and nesting behaviors.

2.5. Puppies

The number of LP and SB was recorded by category.

(a) Liveborn: A total of 225 LP puppies were recorded for the four categories: C1: 63; C2: 71; C3: 54; C4: 37. The neonates, who had a heartbeat and were breathing during the first minute of life, were considered liveborn puppies. The puppies that died once the birth was over, were considered dead during lactation.
(b) Stillbirth: The 47 SB, by the weight categories of their mothers, occurred as follows: C1: 9; C2: 12; C3: 14; C4: 12. The following cases were classified as dead antepartum (i.e., stillbirth Type I): fetuses that died after the birth process began but before the expulsion; those with hemorrhagic and edematous appearance; those with grayish-brown discoloration due to an initial state of mummification; more advanced cases, a clear state of dehydration and fur loss. Those fetuses were excluded. The fetuses classified as dead intrapartum (Type II SB) presented the same appearance as the rest of the litter, except for the absence of breathing and heartbeat.

2.5.1. Blood Physio-Metabolic Profiles

Blood Sampling

A trained veterinarian took blood samples immediately after birth in less than 10 s. An assistant held the puppy in a supine position and exposed the abdominal region; the pup's umbilical cord was carefully grasped to insert the needle (26G) of a tuberculin syringe impregnated with lithium heparin to avoid coagulation and alterations of blood values and immediately obtain 0.3 mL samples of venous blood. All samples were processed by a GEM Premier® critical blood variable analyzer (Instrumentation Laboratory Diagnostics

USA/Italy) to obtain values for the metabolite's glucose (mg/dL) and lactate (mg/dL), blood gases pCO_2 (mmHg) and pO_2 (mmHg), the acid–base balance pH, HCO_3^- (mmol/L), EB (mEq/L), Ca^{2+} (mmol/L), and hematocrit (Htc %). All profiles were tested for each LP and Type II SB pup.

2.5.2. Birth Weight

The weight of each LP and Type II SB puppy was recorded using a digital scale (Salter Weight Tronix Ltd., West Bromwich, UK) after drawing the blood samples.

2.6. Statistical Analyses

Descriptive statistics were obtained for the variables tested following the Origin Version® 9 statistical package procedure. Normality tests were performed for all dependent variables to determine differences among the 4 categories (the dam's weight was the independent variable) regarding the parameters of the blood physio-metabolic profile: pH, pO_2 (mmHg), pCO_2 (mmHg), glucose (mg/dL), Ca^{2+} (mmol/L), lactate (mg/dL), hematocrit (Htc %), HCO_3^- (mmol/L), and EB (mEq/L), as well as the weight of the LP pups (dependent variable). An analysis of variance (ANOVA) was performed with a contrast of means using a Tukey test ($p < 0.05$).

Values were considered as significant at $p < 0.05$.

$$Metabolites_{ijk} = \mu + T_i + CN_i + P_i\ (T_i\ CN_i\ P_i) + e_{ijk}$$

where:

Metabolites = pH, pCO_2, pO_2, glucose, Ca^{2+}, lactate, hematocrit, HCO_3^-, EB;
μ = general mean;
T_i = fixed effect;
CN_i = 1,2,3,4; for the case of SB Type II;
P_i = birth weight;
e = error.

The same statistical model was used to analyze the SB Type II neonates.

2.7. Ethical Note

The studies were performed with privately owned dogs. Each owner gave her/his informed consent before data were gathered. During the study, all dogs were treated following the directives and guidelines of Mexico's Official Norm NOM-062-ZOO-1999 on technical specifications for the production, care, and use of laboratory animals and those related to the field of applied etiological studies [44]. The experimental protocol (code CAMCA.32.18) was approved by the Committee of the Master's Program in Agricultural Sciences of the Universidad Autónoma Metropolitana-Xochimilco, Mexico City.

3. Results

3.1. Weight

Significant statistical differences were observed (see Table 1, mean and standard error of the birth weights of neonates) between the weight of the LP puppies in categories one (189.85 ± 16.50 g; p = 0.01), two (266.84 ± 16.23 g; p = 0.01), and three (374.57 ± 45.18 g; p = 0.0001), with those in C1 having lower weights than those in C2 and C3 (76.99–184.72 g). There were no significant differences in the weight of the LP pups in C3 (374.57 ± 45.18 g) and C4 (381.02 ± 20.24) (p = 0.70).

Table 1 presents data on the weight of the Type II SB pups. The pups in C1 (219.11 ± 23.05 g; p = 0.01) weighed less than those in C2 (297.08 ± 17.62 g; p = 0.001), C3 (419.86 ± 4.57 g; p = 0.001), and C4 (433.75 ± 12.98 g; p = 0.001), but there were no significant differences between the weight of the SB pups in C3 (419.86 ± 4.57 g) and C4 (433.75 ± 12.98 g) (p = 0.40).

The lowest number of SB occurred in C1 with nine (12.5%), compared to C2, C3, and C4, which had 14.45–24.48%.

Table 1. Mean and standard error of the birth weights of the liveborn (LP) puppies and the Type II stillbirth (SB) during the first minute of life, grouped according to the category of the mother.

Category	LP n°	Mean Weight (g) ± SEM	SB Type II n°	Mean Weight (g) ± SEM
C1	63	189.85 ± 2.07 [a]	9	219.11 ± 7.68 [a]
C2	71	266.84 ± 1.92 [b]	12	297.08 ± 5.08 [b]
C3	54	374.57 ± 6.55 [c]	14	419.86 ± 3.62 [c]
C4	37	381.02 ± 3.32 [c]	12	433.75 ± 3.74 [c]

[a,b,c] letters indicate differences in the categories of the mother (weight). Variance analysis ($p < 0.05$); Tukey test for independent samples ($p < 0.05$). SEM—standard error of the mean. Weight of the mothers according to category: C1 (4.00–8.00 kg), C2 (8.10–16.00 kg), C3 (16.10–32.00 kg), C4 (32.10–35.8 kg).

3.2. Blood Physiometabolic Profiles

The effect of the mother's weight on the blood physio-metabolic profiles of LP and SB are shown in Tables 2 and 3, respectively. In general, the weight category of the dam was related to metabolic changes in some of their puppies' critical blood variables.

Table 2. Mean and standard error of the blood physio-metabolic profile of the LP puppies grouped according to the category of the weight of the mother.

	Metabolites	C1 LP = 63 (Mean ± SEM)	C2 LP = 71 (Mean ± SEM)	C3 LP = 54 (Mean ± SEM)	C4 LP = 37 (Mean ± SEM)
Energy metabolism	Lactate (mg/dL)	4.80 ± 0.23 [a]	6.28 ± 0.17 [b]	7.05 ± 0.27 [b,c]	7.27 ± 0.30 [c]
	Glucose (mg/dL)	94.92 ± 1.76 [a]	101.29 ± 2.11 [a]	100.27 ± 3.05 [a]	103.91 ± 3.69 [a]
Calcium and hematocrit	Ca^{2+} (mmol/L)	1.43 ± 0.01 [c]	1.55 ± 0.01 [b]	1.62 ± 0.01 [a]	1.59 ± 0.02 [a,b]
	Hematocrit (%)	44.84 ± 0.55 [c]	48.65 ± 0.41 [b]	50.43 ± 0.39 [a]	49.82 ± 0.55 [a,b]
Acid-base balance	pH	7.38 ± 0.01 [a]	7.33 ± 0.01 [a,b]	7.29 ± 0.01 [b]	7.31 ± 0.02 [a,b]
	pO_2 (mmHg)	17.09 ±0.44 [a]	15.47 ± 0.36 [b]	14.48 ± 0.40 [b]	15.18 ± 0.47 [b]
	pCO_2 (mmHg)	47.69 ± 1.01 [b]	54.42 ± 1.35 [a]	55.98 ± 1.63 [a]	54.78 ± 1.87 [a]
	HCO_3^- (mmol/L)	21.94 ± 0.26 [a]	20.44 ± 0.17 [b]	19.96 ± 0.18 [b]	19.86 ± 0.21 [b]
	EB (mEq/L)	−4.77 ± 0.38 [a]	−5.96 ± 0.36 [a]	−7.03 ± 0.45 [a,b]	−8.82 ± 0.43 [c]

[a,b,c] Letters indicate differences in the category of the mother (weight). Variance analysis ($p < 0.05$); Tukey test for independent samples ($p < 0.05$). SEM—standard error of the mean. LP—number of LP in the samples. Blood samples taken in a maximum of 10 s post-birth. Weight of the mothers according to category: C1 (4.0–8.0 kg), C2 (8.10–16.0 kg), C3 (16.10–32.0 kg), C4 (32.10–35.8 kg).

Table 3. Mean and standard error of the blood physio-metabolic profile of the Type II SB grouped according to the category of weight of the mother.

	Metabolites	C1 Type II SB = 9 (Mean ± SEM)	C2 Type II SB = 12 (Mean ± SEM)	C3 Type II SB = 14 (Mean ± SEM)	C4 Type II SB = 12 (Mean ± SEM)
Energy metabolism	Lactate (mg/dL)	11.44 ± 0.52 ^{b}	12.58 ± 0.31 ^{a,b}	12.5 ± 0.22 ^{a,b}	13.08 ± 0.41 ^{a}
	Glucose (mg/dL)	51.11 ± 3.18 ^{a}	42.58 ± 2.19 ^{a,b}	38.78 ± 3.97 ^{b}	41.91 ± 2.09 ^{a,b}
Calcium and hematocrit	Ca^{2+} (mmol/L)	1.85 ± 0.02 ^{a}	1.89 ± 0.01 ^{a}	1.89 ± 0.02 ^{a}	1.85 ± 0.02 ^{a}
	Hematocrit (%)	59.48 ± 0.52 ^{a}	58.97 ± 0.77 ^{a}	58.33 ± 0.90 ^{a}	58.01 ± 1.15 ^{a}
Acid-base balance	pH	6.79 ± 0.06 ^{a}	6.80 ± 0.04 ^{a}	6.83 ± 0.03 ^{a,b}	6.88 ± 0.02 ^{a}
	pO_2 (mmHg)	9 ± 0.91 ^{a}	6.66 ± 0.63 ^{a,b}	6.21 ± 0.48 ^{a,b}	5.75 ± 0.89 ^{b}
	pCO_2 (mmHg)	81.66 ± 2.08 ^{b}	93.08 ± 2.42 ^{a}	91.78 ± 2.28 ^{a}	94.66 ± 1.98 ^{a}
	HCO_3^- (mmol/L)	17.63 ± 0.25 ^{a}	18.80 ± 0.57 ^{a}	18.37 ± 0.44 ^{a}	18.45 ± 0.46 ^{a}
	EB (mEq/L)	−14.42 ± 0.72 ^{a}	−15.33 ± 0.48 ^{a}	−14.26 ± 0.39 ^{a}	−14.38 ± 0.86 ^{a}

^{a,b} Letters indicate differences in the category of the mother (weight). Variance analysis ($p < 0.05$); Tukey test for independent samples ($p < 0.05$). SEM—standard error of the mean. LP—number of LP in the samples. SB—Stillbirth. Blood samples taken in a maximum of 10 s post-birth. Weight of the mothers according to category: C1 (4.0–8.0 kg), C2 (8.10–16.0 kg), C3 (16.10–32.0 kg), C4 (32.10–35.8 kg).

3.2.1. Energy Metabolism

In terms of the level of blood lactate (see Table 2), there was a significant increase of 9.26–14.33% between the LP in C1 (4.80 ± 0.23 mg/dL, p = 0.001) and those in C2 (6.28 ± 0.17 mg/dL, p = 0.001), C3 (7.05 ± 0.27 mg/dL, p = 0.01), and C4 (7.27 ± 0.30 mg/dL, p = 0.01). The LP in C4 had the largest increase in blood lactate concentrations compared to C1 and C2 ($p < 0.05$), though there was no significant difference in lactate values between C2 and C3 (6.28 ± 0.17 mg/dL and 7.05 ± 0.27 mg/dL, respectively) (p = 0.07) (Table 2).

The Type II SB pups in C4 (13.08 ± 0.41 mg/dL) showed a larger increase in lactate than C1 (11.44 ± 0.52) (p = 0.02), but there were no significant differences between C2 and C3 (12.58 ± 0.31 and 12.5 ± 0.22, respectively) (p = 0.99) (Table 3).

In terms of blood glucose levels for the LP, no significant differences among the categories were found ($p > 0.05$) (Table 2). In contrast, the Type II SB pups in C1 (51.11 ± 3.18 mg/dL) had higher glucose levels than those in C3 (38.78 ± 3.97) (p = 0.04). There were no significant differences between C2 and C4 (42.58 ± 2.19 mg/dL and 41.91 ± 2.09 mg/dL, respectively) (p = 0.99) (Table 3).

3.2.2. Calcium and Hematocrit

Regarding the metabolite Ca^{2+}, a statistically significant decrease was found in the values of the LP in C1 (1.43 ± 0.01 mmol/L) compared to C2 (1.55 ± 0.01 mmol/L, p = 0.001), C3 (1.62 ± 0.01 mmol/L, p = 0.01), and C4 (1.59 ± 0.02 mmol/L, p = 0.01), but no significant differences in blood Ca^{2+} concentrations were seen among C2, C3, and C4 ($p > 0.05$) (Table 2). For the Type II SB, Ca^{2+} levels showed no significant differences among the categories ($p > 0.05$) (Table 3).

Hematocrit values showed that the LP from the C2 (48.65 ± 0.41, 8.5%; p = 0.001), C3 (50.43 ± 0.39, 12.46%; p = 0.01), and C4 (49.82 ± 0.55, 11.1%; p = 0.01) dams all had higher percentages of hematocrit than those in C1 (44.84 ± 0.55) ($p < 0.05$). However, no significant differences were seen among the LP in C2, C3, and C4 ($p > 0.05$) (Table 3). Regarding the Type II SB, there were no significant differences among the categories for this value ($p > 0.05$) (Table 3).

3.2.3. Acid–Base Balance

Observations showed that blood pH had a lower imbalance in the LP from the C1 dams (7.38 ± 0.01; $p < 0.05$), compared to C3 (7.29 ± 0.01; $p = 0.01$) and C4 (7.31 ± 0.02; $p = 0.04$). While pH clearly decreased, there were no significant differences in pH values among categories C2, C3, and C4 ($p < 0.05$) (Table 2). Similarly, for the Type II SB, there were no significant differences among the categories in terms of pH ($p > 0.05$) (Table 3).

Regarding pO_2, results showed that the LP from three groups—C2 (15.47 ± 0.36 mmHg, $p = 0.01$), C3 (14.48 ± 0.40 mmHg, $p = 0.001$), and C4 (15.18 ± 0.47 mmHg, $p = 0.01$)—had significant decreases compared to C1 (17.09 ± 0.44 mmHg) ($p < 0.05$), but there were no significant differences in the pO_2 values among C2, C3, and C4 ($p > 0.05$) (Table 3). For blood pO_2 values, we found that the Type II SB in C4 (5.75 ± 0.89 mmHg) had a significant decrease compared to C1 (9 ± 0.91 mmHg) ($p = 0.02$), but the pO_2 values between C2 and C3 did not differ significantly (6.66 ± 0.63 mmHg, $p = 0.15$, and 6.21 ± 0.48 mmHg, $p = 0.053$, respectively) (Table 3).

Regarding pCO_2, a statistically significant increase was observed in the LP in C2 (54.42 ± 1.35 mmHg, $p = 0.01$), C3 (55.98 ± 1.63 mmHg, $p = 0.001$), and C4 (54.78 ± 1.87 mmHg, $p = 0.01$), compared to C1 (47.69 ± 1.01 mmHg). In contrast, no significant differences were found in the pCO_2 values of C2, C3, and C4 ($p > 0.05$) (Table 2). Upon observing the blood pCO2 values in the Type II SB, a significant increase was found in groups C2 (93.08 ± 2.42 mmHg, $p = 0.01$), C3 (91.78 ± 2.28 mmHg, $p = 0.01$), and C4 (94.66 ± 1.9 mmHg, $p = 0.01$) compared to C1 (1.66 ± 2.08 mmHg), but no significant differences were found in the pCO_2 values from groups C2, C3, and C4 ($p > 0.05$) (Table 3).

A significant decrease in HCO_3^- levels was seen in the LP in C2 (20.44 ± 0.17 mmol/L, $p = 0.001$), C3 (19.96 ± 0.18 mmol/L, $p = 0.01$), and C4 (19.86 ± 0.21 mmol/L, $p = 0.001$) compared to C1 (21.94 ± 0.26 mmol/L) ($p < 0.05$), but upon comparing the metabolite values for C2, C3, and C4, no significant differences were found ($p > 0.05$) (Table 2). Regarding the SB, no differences were observed among the categories ($p > 0.05$) (Table 3).

Observations showed that EB increased significantly in the LP in C4 (−8.82 ± 0.43 mEq/L) compared to the values for C1 (−4.77 ± 0.38 mEq/L, $p = 0.01$), C2 (−5.96 ± 0.36 mEq/L, $p = 0.001$), and C3 (−7.03 ± 0.45 mEq/L, $p = 0.01$). A similar result emerged when we compared the values for C1 and C3 ($p = 0.001$), though no significant differences were found between C2 and C3 ($p = 0.11$) (Table 2). Regarding the Type II SB, no significant differences were found among the categories ($p > 0.05$) (Table 3).

4. Discussion

4.1. Weight

It was proposed that the weight of puppies at birth can be influenced by diverse factors, as occurs in other mammals. These include the duration of pregnancy [27,45,46], restrictions on intrauterine growth [45,47,48], the mother's nutritional status [27], breed [49], and the weight and size of the placenta [50]. However, the results of our study suggest that the mother's weight before giving birth exerts an effect on the weight of the newborn puppies since this varied significantly among the four categories tested. A broad weight range was observed that might be attributable to this species' extensive morphological variability characteristic [51]. The recorded weights of the 272 puppies born and classified according to the weight of the dam (based on FCI guidelines, [43]) showed a mean from 204.48 to 407.39 g (157–453 g), with a mean variation of 77.46 to 202.91 g. We believe that the weight of the neonates reflected the mother's weight because our study did not consider breed as a variable. A study by Vassalo et al. [6] observed a similar effect, the mothers' body weight influenced the weight of puppies born by eutocic births and cesarean section. While it is true that the dam's nutritional status can affect the weight of the puppies—as occurs in humans [52]—this variable was not controlled in this study.

One important finding involves the weight of the SB, as this was always higher than that of the LP in all four cases (C1–C4). These differences mean 29.16–52.77 g, but categories C3 and C4 had both the highest mean weights (419.86 and 433.79 g, respectively) and the highest mortality rates (C3 = 20.58%, C4 = 24.58%). On one side, and such as other species, including humans, swine, and bovines, low birth weight is considered an important risk factor for neonatal mortality [34,53,54]. Reports on dogs affirm that low birth weight is strongly related to mortality. In this regard, Groppetti et al. [55] and Mila et al. [56] pointed out that there is a twelve-fold higher mortality risk for the lightest puppies than those with normal weight. Though low birth weight has been deemed a disadvantageous condition for neonatal survival [3,55,57,58] and has been associated with a higher risk of fetal death, our study did not reveal signs of this because the heaviest puppies presented a higher risk of fetal death than those with the lowest weight (12.5 vs. 24.58% mortality). On the other side, previous results indicate that higher birth weights reduce postnatal mortality but increase the rate of intrapartum mortality due to the difficulties of birth caused by cephalopelvic disproportion and prolonged labor that can cause hypoxia or death [28,59,60]. The mortality rate in this study was 17.27%, counting only the pups that died intrapartum. While it is true that the cause of perinatal mortality in dogs is multifactorial, the mother's weight must be considered a risk factor due to its impact on the weight of type II SB puppies, physiological alterations, and the acid–base imbalance present in puppies born in natural births.

4.2. Physiometabolic Profiles

Fetuses commonly suffer intermittent periods of light hypoxia due to uterine contractions and the mechanical pressure inherent to the birth process [61]. Vassalo et al. [6] affirmed that a state of fetal hypoxia during the perinatal period is common in newborn puppies. We measured the physiological and metabolic changes that LP experienced during eutocic births, including increases in blood levels of pCO_2 and lactate of 17.38% and 51.45%, respectively, and decreases in blood pH of 1.22%, and levels of pO_2 (15.27%) and bicarbonate HCO_3^- (9.48%), with high EB (45.91%), all due to hypercapnia (an indicator of respiratory acidosis) as the main factor. The most evident alterations occurred in the LP groups with the heaviest dams (C2, C3, C4). These alterations in gases and blood metabolites indicated respiratory and metabolic acidosis (mixed acidosis) resulting from intermittent asphyxia in utero during natural birth. Compensatory alkalosis began as a response to the respiratory and metabolic acidosis in the LP in all groups (C1–C4) due to hypoxia. This explains why pH did not decrease drastically [62,63]. Hypoxia-induced stress increases circulating epinephrine that breaks down muscular glycogen; thus, increasing lactate concentrations [64–66]. The above slows metabolism and triggers delayed anaerobiosis, a mechanism known as a tolerance to fetal hypoxia [67].

Regarding the metabolite glucose, no significant differences were observed in the LP during the first minute of life in any of the four categories, since measurements were in a range of 94.92–103.91 mg/dL. Mila et al. [17] reported a mean plasma glucose concentration of 97 mg/dL between 10 min and 8 h postpartum. The blood glucose level of puppies in the first 24 h postpartum was established in a range of 88–133 mg/dL [68,69]. Adequate energy reserves are extremely important for neonatal survival and resistance to adverse climatic conditions [70]. However, during the first hours of life, a decrease in glucose concentrations may be seen in diverse species because the glucose supply is interrupted abruptly during birth. This decrease is associated with the rapid exhaustion of hepatic glycogen. Hypoglycemia increases blood glucagon, cortisol, and catecholamine levels, leading to gluconeogenesis, lipolysis, glycogenolysis, and the consumption of ketone bodies. Ingesting colostrum post-birth increases and maintains glucose levels. In humans, for example, values below 50 mg/dL have been observed, though these may increase to 81 mg/dL during daytime [71,72]. A similar situation has been seen in foals [73]. On another point, an increase in blood glucose in newborn piglets can be considered an accurate indicator of neonatal distress because it shows their incapacity to regulate, or compensate for, the physiological processes during birth [13,37,74]. Mota-Rojas et al. [75] mention that high glucose concentrations in piglets are a sign of a short episode of asphyxia compared to those that manage to maintain their energy reserves. Therefore, lower glucose levels are associated with a more extended period of asphyxia and higher a consumption of energy reserves. It is important to mention that a prolonged or intermittent asphyxia in utero during birth does not necessarily lead to intrapartum stillbirth. However, these conditions can weaken newborns and reduce their capacity to adapt to extrauterine life, as documented in piglet studies [36]. The events that occur during an acute process of asphyxiation—such as metabolic acidosis and hypoxia—impact the welfare of newborns and their postnatal development.

Birth weight was reported as a risk factor for intrapartum hypoxia because newborns with a low birth weight are more likely to suffer oxygen restriction and the secondary effects of hypoxemia [76,77]. Present findings, however, indicate otherwise. The blood samples collected from the umbilical cords for the gas and metabolic analyses of the intrapartum SB indicated that the fetuses showed signs of severe metabolic acidosis moments before birth due to the low pH (range: 6.79–6.88), increased pCO_2 levels up to double those registered in the LP (94.66 vs. 47.69 mmHg), lactate values as many as four times higher than in the LP (13.08 vs. 4.80 mg/dL), EB in a range of -14.26 to -15.33, and a decrease in pO_2 (6.22%) and bicarbonate HCO_3^- levels (6.97%). We also observed that the Type II SB showed a significant decrease in plasma glucose concentrations in every group (C1–C4), with a range of 38.78–51.11 mg/dL. This result coincides with the values < 40 mg/dL reported by Lawler [78], related to hypoglycemia, which indicate a depletion of the newborns' energy reserves, as a by-product of a previous hypoxic process [7,78] accompanied by a delay in eliminating excess liquid from the lungs, a decrease in uterine blood circulation, deficient gas exchange, and an alteration of energy metabolism [16]. We, therefore, infer that prolonged uterine contractions without fetal expulsion, caused by the prolonged birth process characteristic of this species, trigger hypoxia by increasing anaerobic glycogenolysis and the development of metabolic acidosis, as has been seen in piglets [75,79,80], humans [81], foals [82], and domesticated animals, including buffaloes [83,84]. These are conditions that reduce vitality and increase mortality [85].

Fetal asphyxia, defined as a condition of hypoxemia with hypercapnia and acidosis, caused the death of the pups at birth in this study. Fetal asphyxia caused the death of 17.27% of the puppies in this study, but the highest percentages of intrapartum deaths were seen in C3 and C4, which together represented 55.31% of the mortality of the pups from natural births.

Acid–Base Balance

In LP, a pH below the reference values was observed (7.35–7.45) in C2, C3, and C4, but this parameter on its own does not permit measuring the accumulated exposure to hypoxia because it is expressed logarithmically [39]. It does, however, establish the existence of acidosis in newborns and reflects fetal hypoxic stress that occurred during birth. Hence, the relation among pH, bicarbonate, and pCO_2 indicates a process of metabolic acidosis [86]. In addition, the concentration of excess base (EB) must be determined, as this is an indicator of a linear tendency that determines the accumulation of acidosis after being adjusted for variations in pCO_2 [87].

The intense, constant uterine contractions necessary for fetal expulsion can compress the umbilical cord, drastically decreasing both placental and umbilical blood circulation [88]. The decrease in pO_2 and the increase in pCO_2 combined to lower the pH, but bicarbonate in the plasma mitigated this imbalance. As a result, the high concentration of bicarbonate generated a mixed acidosis that exacerbated the increase in ionized calcium and the decrease in protein-bonded calcium [89], producing hypocalcemia at birth. Andres et al. [25] indicate that delayed breathing and metabolic acidosis are related to mortality at birth.

During pregnancy, fetuses depend on their mother for gas exchange and correct oxygenation through the placenta. This exchange is determined by the size of the fetuses, blood gas concentrations in the mother, and the latter's capacity for transfer and transport. Thus, modifications to any of these parameters can generate a state of hypoxia and subsequent disruptions of the acid–base balance, such as metabolic acidosis [61]. It is believed that the fetus' capacity to withstand birth stress and welfare depends on both its condition at birth and the birthing process itself (duration, number of contractions, fetus thermoregulation, physiological, and metabolic changes. In addition, the newborn has to make several adjustments to adapt to extrauterine life, such as maintaining normoglycemia, thermoregulation, etc.) [18,90–95].

5. Conclusions

The results of this study allowed us to identify that the weight of dams before birth determines the weight of the puppies at birth, though there is a wide range in birth weights due to the ample morphological variability characteristic of this species. We observed that the puppies born from bitches in categories three and four had the highest birth weights (397.21 and 407.39 g, respectively), with a difference greater than 100 g from the pups in C1 and C2. The LP puppies in category 1—the lightest ones (189.85 g)—had fewer physio-metabolic alterations during the first minute of life; while the litters that weighed 16.1–35.8 kg showed more physio-metabolic alterations, which impacted their adaption to extrauterine life and caused respiratory and metabolic acidosis and hypocalcemia, all of which contributed to the total mortality of 17% as a consequence of intrapartum hypoxia.

Furthermore, all the Type II SB were heavier (C1 > 29.26 g; C2 > 30.24 g; C3 > 45.29 g; C4 > 52.73 g) than the LP born in the same group. The above suggests that puppies with higher weights experience greater difficulty during expulsion through the birth channel, which increases the risk of suffering an acute process of intrauterine asphyxia that affects their chances of survival. The highest mortality rates for the heavier categories (C3 and C4) were 20.58% and 24.48%, respectively. Therefore, this study established the mother's weight (16.1–35.8 kg) as a risk factor for fetal asphyxia in eutocic births and, consequently, a high percentage of intrapartum mortality. New studies will allow the integrated evaluation of other variables as risk factors related to in utero asphyxia in eutocic births and their influence on postpartum survival.

Author Contributions: Conceptualization, D.M.-R. and A.G.; investigation, B.R.-S., D.M.-R., P.M.-M., A.O.-H., I.H.-Á., J.M.-B. and J.S.-M.; writing—original draft preparation, D.M.-R., B.R.-S., A.O., C.M., A.G. and J.M.-B.; writing—review and editing, D.M.-R., A.O.-H., B.R.-S., J.M.-B., P.M.-M., J.S.-M., I.H.-Á. and B.R.-S.; project administration, D.M.-R., A.O., C.M. and A.G. All authors have read and agreed to the published version of the manuscript.

Funding: This research received no external funding.

Institutional Review Board Statement: The study was conducted according to the guidelines of the Declaration of Helsinki, and the experimental protocol was approved by the Committee of the Master's Program in Agricultural Sciences (protocol code CAMCA.32.18) of the Universidad Autónoma Metropolitana-Xochimilco campus, Mexico City.

Conflicts of Interest: The authors declare that they have no conflict of interest.

Abbreviations/Nomenclature

pO_2	partial oxygen saturation
O_2	oxygen
EB	excess base
HCO_3^-	bicarbonate
Ca^{2+}	calcio
Htc	hematocrit

References

1. Veronesi, M.C. Assessment of canine neonatal viability—The Apgar score. *Reprod. Domest. Anim.* **2016**, *51*, 46–50. [CrossRef] [PubMed]
2. Kirkden, R.D.; Broom, D.M.; Andersen, I.L. Invited review: Piglet mortality: Management solutions. *J. Anim. Sci.* **2013**, *91*, 3361–3389. [CrossRef] [PubMed]
3. Tønnessen, R.; Borge, K.S.; Nødtvedt, A.; Indrebø, A. Canine perinatal mortality: A cohort study of 224 breeds. *Theriogenology* **2012**, *77*, 1788–1801. [CrossRef] [PubMed]
4. Veronesi, M.C.; Panzani, S.; Faustini, M.; Rota, A. An Apgar scoring system for routine assessment of newborn puppy viability and short-term survival prognosis. *Theriogenology* **2009**, *72*, 401–407. [CrossRef]
5. Gill, M.A. Perinatal and late neonatal mortality in the dog. *J. Am. Vet. Med. Assoc.* **2001**, *230*, 1463–1464.
6. Vassalo, F.G.; Simões, C.R.B.; Sudano, M.J.; Prestes, N.C.; Lopes, M.D.; Chiacchio, S.B.; Lourenço, M.L.G. Topics in the routine assessment of newborn puppy viability. *Top. Companion Anim. Med.* **2015**, *30*, 16–21. [CrossRef]
7. Münnich, A.; Küchenmeister, U. Causes, diagnosis and therapy of common diseases in neonatal puppies in the first days of life: Cornerstones of practical approach. *Reprod. Domest. Anim.* **2014**, *49*, 64–74. [CrossRef] [PubMed]
8. Grundy, S.A. Clinically relevant physiology of the neonate. *Vet. Clin. N. Am. Small Anim. Pract.* **2006**, *36*, 443–459. [CrossRef] [PubMed]
9. Manani, M.; Jegatheesan, P.; DeSandre, G.; Song, D.; Showalter, L.; Govindaswami, B. Elimination of admission hypothermia in preterm very low-birth-weight infants by standardization of delivery room management. *Perm. J.* **2013**, *17*, 8–13. [CrossRef]
10. Cavaliere, T.A. From fetus to neonate: A sensational journey. *Newborn Infant Nurs. Rev.* **2016**, *16*, 43–47. [CrossRef]
11. Mota-Rojas, D.; Villanueva-García, D.; Hernández, G.R.; Roldan, S.P.; Martínez-Rodríguez, R.; Mora-Medina, P.; Trujillo-Ortega, M.E. Assessment of the vitality of the newborn: An overview. *Sci. Res. Essays* **2012**, *7*. [CrossRef]
12. Mota-Rojas, D.; Martinez-Burnes, J.; Villanueva-Garcia, D.; Roldan, P.; Trujillo-Ortega, M.E.; Orozco, H.; Bonilla, H.; Lopez-Mayagoitia, A. Animal welfare in the newborn piglet: A review. *Vet. Med.* **2012**, *57*, 338–349. [CrossRef]
13. Mota-Rojas, D.; Fierro, R.; Roldan, P.; Orozco, H.; González, M.; Martínez-Rodríguez, R.; García-Herrera, R.; Mora-Medina, P.; Flores-Peinado, S.; Sánchez, M.; et al. Outcomes of gestation length in relation to farrowing performance in sows and daily weight gain and metabolic profiles in piglets. *Anim. Prod. Sci.* **2015**. [CrossRef]
14. Nuñez, A.; Benavente, I.; Blanco, D.; Boix, H.; Cabañas, F.; Chaffanel, M.; Fernández-Colomer, B.; Fernández-Lorenzo, J.R.; Loureiro, B.; Moral, M.T.; et al. Estrés oxidativo en la asfixia perinatal y la encefalopatía hipóxico-isquémica. *An. Pediatría* **2018**. [CrossRef]
15. Kuttan, K.V.; Joseph, M.; Simon, S.; Ghosh, K.N.A.; Rajan, A. Effect of intrapartum fetal stress associated with obstetrical interventions on viability and survivability of canine neonates. *Vet. World* **2016**, *9*, 1485–1488. [CrossRef]
16. Hillman, N.H.; Kallapur, S.G.; Jobe, A.H. Physiology of transition from intrauterine to extrauterine life. *Clin. Perinatol.* **2012**, *39*, 769–783. [CrossRef] [PubMed]
17. Mila, H.; Grellet, A.; Delebarre, M.; Mariani, C.; Feugier, A.; Chastant-Maillard, S. Monitoring of the newborn dog and prediction of neonatal mortality. *Prev. Vet. Med.* **2017**, *143*, 11–20. [CrossRef]
18. Lezama-García, K.; Mariti, C.; Mota-Rojas, D.; Martínez-Burnes, J.; Barrios-García, H.; Gazzano, A. Maternal behaviour in domestic dogs. *Int. J. Vet. Sci. Med.* **2019**, *7*, 20–30. [CrossRef]
19. Reyes-Sotelo, B.; Mota-Rojas, D.; Martínez-Burnes, J.; Olmos-Hernández, A.; Hernández-Ávalos, I.; José, N.; Casas-Alvarado, A.; Gómez, J.; Mora-Medina, P. Thermal homeostasis in the newborn puppy: Behavioral and physiological responses. *J. Anim. Behav. Biometeorol.* **2021**, *9*. [CrossRef]
20. Mota-Rojas, D.; Orihuela, A.; Strappini, A.; Villanueva-García, D.; Napolitano, F.; Mora-Medina, P.; Barrios-García, H.B.; Herrera, Y.; Lavalle, E.; Martínez-Burnes, J. Consumption of maternal placenta in humans and nonhuman mammals: Beneficial and adverse effects. *Animals* **2020**, *10*, 2398. [CrossRef]
21. Ogi, A.; Mariti, C.; Pirrone, F.; Baragli, P.; Gazzano, A. The influence of oxytocin on maternal care in lactating dogs. *Animals* **2021**, *11*, 1130. [CrossRef]
22. Massip, A. Relationship between pH, plasma, cortisol and glucose concentrations in the calf at birth. *Br. Vet. J.* **1980**, *136*, 597–601. [CrossRef]
23. Wiberg, N.; Källén, K.; Olofsson, P. Physiological development of a mixed metabolic and respiratory umbilical cord blood acidemia with advancing gestational age. *Early Hum. Dev.* **2006**. [CrossRef]
24. Bleul, U.; Lejeune, B.; Schwantag, S.; Kähn, W. Blood gas and acid-base analysis of arterial blood in 57 newborn calves. *Vet. Rec.* **2007**. [CrossRef] [PubMed]
25. Andres, R.L.; Saade, G.; Gilstrap, L.C.; Wilkins, I.; Witlin, A.; Zlatnik, F.; Hankins, G.V. Association between umbilical blood gas parameters and neonatal morbidity and death in neonates with pathologic fetal acidemia. *Am. J. Obstet. Gynecol.* **1999**, *181*, 867–871. [CrossRef]
26. Le Cozler, Y.; Guyomarc'h, C.; Pichodo, X.; Quinio, P.Y.; Pellois, H. Factors associated with stillborn and mummified piglets in high-prolific sows. *Anim. Res.* **2002**. [CrossRef]
27. Lucia, T.; Corrêa, M.N.; Deschamps, J.C.; Bianchi, I.; Donin, M.A.; Machado, A.C.; Meincke, W.; Matheus, J.E.M. Risk factors for stillbirths in two swine farms in the south of Brazil. *Prev. Vet. Med.* **2002**, *53*, 285–292. [CrossRef]

28. Vanderhaeghe, C.; Dewulf, J.; de Kruif, A.; Maes, D. Non-infectious factors associated with stillbirth in pigs: A review. *Anim. Reprod. Sci.* **2013**, *139*, 76–88. [CrossRef] [PubMed]
29. Johnson, C.A. Pregnancy management in the bitch. *Theriogenology* **2008**. [CrossRef]
30. Mila, H.; Grellet, A.; Chanstant-Maillard, S. Pronostic value or birth weight and early weight gain on neonatal and pediatric mortality: A longitudinal study on 870 puppies. In Proceedings of the Program and Presented at the 7th International Symposium on Canine and feline Reproduction 2012, Wistler, BC, Canada, 26–29 July 2012; pp. 163–164.
31. Crissiuma, A.L.; Juppa, C.J., Jr.; Mendes de Almeida, F.; Gershony, L.C.; Labarthe, N.V. Influence of the order of birth on blood gasometry parameters in the fetal neonatal transitional period of dogs born by elective caesarean parturition. *J. Appl. Res. Vet. Med.* **2010**, *8*, 7–15.
32. Mota-Rojas, D.; Orozco, H.; Alonso-Spilsbury, M.; Villanueva-García, D.; Martinez-Burnes, J.; Lopez-Mayagoitia, A.; Gonzalez, M.; Trujillo, M.E.; Ramirez, R. Asfixia perinatal en el bebe y neonato porcino. In *Perinatología Animal: Enfoques Clínicos Experimentales*, 2nd ed.; BM Editores: Mexico City, Mexico, 2008; pp. 287–304. ISBN 9789709502411.
33. Martínez-Rodríguez, R.; Mota-Rojas, D.; Trujillo-Ortega, M.E.; Orozco, H.; Hernández, R.; Mora-Medina, P.; Alonso-Spilsbury, M.; Rosales-Torres, A.; Ramírez-Necoechea, R. Physiological response to hypoxia in piglets of different birth weight. *Ital. J. Anim. Sci.* **2011**, *2011*, 250–253. [CrossRef]
34. Mota-Rojas, D.; López, A.; Martínez-Burnes, J.; Muns, R.; Villanueva-García, D.; Mora-Medina, P.; González-Lozano, M.; Olmos-Hernández, A.; Ramírez-Necoechea, R. Is vitality assessment important in neonatal animals? *CAB Rev. Perspect. Agric. Vet. Sci. Nutr. Nat. Resour.* **2018**, *13*. [CrossRef]
35. Proulx, J. Respiratory monitoring: Arterial blood gas analysis, pulse oximetry, and end-tidal carbon dioxide analysis. *Clin. Tech. Small Anim. Pract.* **1999**, *14*, 227–230. [CrossRef]
36. Trujillo-Ortega, M.E.; Mota-Rojas, D.; Olmos-Hernández, A.; González, M.; Ramírez-Necoechea, R. A study of piglets born by spontaneous parturition under uncontrolled conditions: Could this be a naturalistic model for the study of intrapartum asphyxia? *Acta Biomed.* **2007**, *78*, 29–35. [PubMed]
37. Mota-Rojas, D.; Orozco, H.; Villanueva-García, D.; Hernández-González, R.; Roldan, P.; Trujillo-Ortega, M.E. Foetal and neonatal energy metabolism in pigs and humans: A review. *Veterninarni Med.* **2011**, *56*, 215–225. [CrossRef]
38. Villanueva-García, D.; Mota-Rojas, D.; González, M.; Olmos-Hernández, S.A.; Orozco, H.; Sánchez, P. Importancia de la gasometría en perinatología. In *Perinatología Animal: Enfoques Clínicos Experimentales*, 2nd ed.; BM Editores: Mexico City, Mexico, 2008; pp. 183–192. ISBN 970-95024-0-9.
39. Armstrong, L.; Stenson, B.J. In the assessment of the newborn. *Arch. Dis. Child. Fetal Neonatal Ed.* **2007**, *92*, 430–434. [CrossRef]
40. Blickstein, I.; Green, T. Umbilical cord blood gases. *Clin. Perinatol.* **2007**, *34*, 451–459. [CrossRef]
41. Apgar, V. A proposal for a new method of evaluation of the newborn infant. *Anesth. Analg.* **2015**, *120*, 1056–1059. [CrossRef]
42. World Small Animal Veterinary Association Global Nutritional Assesment Guidelines. Available online: http://wsava.org/wp-content/uploads/2020/01/Global-Nutritional-Assesment-Guidelines-Spanish.pdf (accessed on 21 January 2021).
43. Federation Cynologique Internationale (FCI). Available online: http://www.fci.be (accessed on 18 May 2021).
44. Sherwin, C.M.; Christiansen, S.B.; Duncan, I.J.; Erhard, H.W.; Lay, D.C.; Mench, J.A.; O'Connor, C.E.; Carol Petherick, J. Guidelines for the ethical use of animals in applied ethology studies. *Appl. Anim. Behav. Sci.* **2003**, *81*, 291–305. [CrossRef]
45. Meyer, K.M.; Koch, J.M.; Ramadoss, J.; Kling, P.J.; Magness, R.R. Ovine surgical model of uterine space restriction: Interactive effects of uterine anomalies and multifetal gestations on fetal and placental growth. *Biol. Reprod.* **2010**, *83*, 799–806. [CrossRef]
46. Rydhmer, L.; Lundeheim, N.; Canario, L. Genetic correlations between gestation length, piglet survival and early growth. *Livest Sci.* **2008**, *115*, 287–293. [CrossRef]
47. Bautista, A.; Rödel, H.G.; Monclús, R.; Juárez-Romero, M.; Cruz-Sánchez, E.; Martínez-Gómez, M.; Hudson, R. Intrauterine position as a predictor of postnatal growth and survival in the rabbit. *Physiol. Behav.* **2015**, *138*, 101–106. [CrossRef]
48. Darby, J.R.T.; Varcoe, T.J.; Orgeig, S.; Morrison, J.L. Cardiorespiratory consequences of intrauterine growth restriction: Influence of timing, severity and duration of hypoxaemia. *Theriogenology* **2020**, *150*, 84–95. [CrossRef] [PubMed]
49. Groppetti, D.; Pecile, A.; Palestrini, C.; Marelli, S.P.; Boracchi, P. A national census of birthweight in purebred dogs in Italy. *Animals* **2017**, *7*, 43. [CrossRef]
50. Tesi, M.; Miragliotta, V.; Scala, L.; Aronica, E.; Lazzarini, G.; Fanelli, D.; Abramo, F.; Rota, A. Theriogenology relationship between placental characteristics and puppies' birth weight in toy and small sized dog breeds. *Theriogenology* **2020**, *141*, 1–8. [CrossRef]
51. Mugnier, A.; Mila, H.; Guiraud, F.; Brévaux, J.; Lecarpentier, M.; Martinez, C.; Mariani, C.; Adib-Lesaux, A.; Chastant-Maillard, S.; Saegerman, C.; et al. Birth weight as a risk factor for neonatal mortality: Breed-specific approach to identify at-risk puppies. *Prev. Vet. Med.* **2019**, *171*. [CrossRef]
52. Vasudevan, C.; Renfrew, M.; McGuire, W. Fetal and perinatal consequences of maternal obesity. *Arch. Dis. Child. Fetal Neonatal Ed.* **2011**, *96*, F378–F382. [CrossRef]
53. Merton, J.S.; van Wagtendonk-de Leeuw, A.M.; den Dass, J.H. Factors affecting birthweight of IVP calves. *Progress. Cattle* **2014**, *49*, 293.
54. Wu, G.; Bazer, F.W.; Wallace, J.M.; Spencer, T.E. Board-invited review: Intrauterine growth retardation: Implications for the animal sciences. *J. Anim. Sci.* **2006**, *84*, 2316–2337. [CrossRef] [PubMed]
55. Groppetti, D.; Ravasio, G.; Bronzo, V.; Pecile, A. The role of birth weight on litter size and mortality within 24h of life in purebred dogs: What aspects are involved? *Anim. Reprod. Sci.* **2015**. [CrossRef] [PubMed]

56. Mila, H.; Grellet, A.; Feugier, A.; Chastant-Maillard, S. Differential impact of birth weight and early growth on neonatal mortality in puppies 1, 2. *J. Anim. Sci.* **2015**, *93*, 4436–4442. [CrossRef]
57. Herpin, P.; Damon, M.; Le Dividich, J. Development of thermoregulation and neonatal survival in pigs. *Livest. Prod. Sci.* **2002**. [CrossRef]
58. Islas-Fabila, P.; Mota-Rojas, D.; Martínez-Burnes, J.; Mora-Medina, P.; González-Lozano, M.; Roldán, P.; Greenwell, V.; González, M.; Vega, X.; Gregorio, H. Physiological and metabolic responses in newborn piglets associated with the birth order. *Anim. Reprod. Sci.* **2018**, *197*, 247–256. [CrossRef]
59. Cornelius, A.J.; Moxon, R.; Russenberger, J.; Havlena, B.; Cheong, S.H. Identifying risk factors for canine dystocia and stillbirths. *Theriogenology* **2019**, *128*, 201–206. [CrossRef]
60. Dolf, G.; Gaillard, C.; Russenberger, J.; Moseley, L.; Schelling, C. Factors contributing to the decision to perform a cesarean section in Labrador retrievers. *BMC Vet. Res.* **2018**, *14*, 1–9. [CrossRef]
61. Yli, B.M.; Kjellmer, I. Pathophysiology of foetal oxygenation and cell damage during labour. *Best Pract. Res. Clin. Obstet. Gynaecol.* **2016**, *30*, 9–21. [CrossRef]
62. Lúcio, C.F.; Silva, L.C.G.; Rodrigues, J.A.; Veiga, G.A.L.; Vannucchi, C.I. Acid-base changes in canine neonates following normal birth or dystocia. *Reprod. Domest. Anim.* **2009**, *44*, 208–210. [CrossRef] [PubMed]
63. Vivan, M.C.M. Correlation between Serum Lactate and Neurological and Cardiorrespiratory Status in Neonate Dogs Born by Normal Birth or Cesarean Section under Inhalatory Anesthesia. Ph.D. Dissertation, University of Sao Paulo State, Sao Paulo, Brazil, 2010.
64. Gregory, N.G. *Animal Welfare and Meat Science*; CABI: Wallingford, UK, 1998; ISBN 085199296X.
65. Groppetti, D.; Pecile, A.; Del Carro, A.P.; Copley, K.; Minero, M.; Cremonesi, F. Evaluation of newborn canine viability by means of umbilical vein lactate measurement, apgar score and uterine tocodynamometry. *Theriogenology* **2010**, *74*, 1187–1196. [CrossRef] [PubMed]
66. Piquard, F.; Schaefer, A.; Dallenbach, P.; Haberey, P. Is fetal acidosis in the human fetus maternogenic during labor? A reanalysis. *Am. J. Physiol.* **1991**, *261*, R1294–R1299. [CrossRef] [PubMed]
67. Singer, D. Neonatal tolerance to hypoxia: A comparative-physiological approach. *Comp. Biochem. Physiol. Part A Mol. Integr. Physiol.* **1999**, *123*, 221–234. [CrossRef]
68. Ishii, T.; Hori, H.; Ishigami, M.; Mizuguchi, H.; Watanabe, D. Background data for hematological and blood chemical examinations in juvenile beagles. *Exp. Anim.* **2013**, *62*, 1–7. [CrossRef]
69. Rosset, E.; Rannou, B.; Casseleux, G.; Chalvet-Monfray, K.; Buff, S. Age-related changes in biochemical and hematologic variables in Borzoi and Beagle puppies from birth to 8 weeks. *Vet. Clin. Pathol.* **2012**. [CrossRef] [PubMed]
70. Nowak, R.; Poindron, P. From birth to colostrum: Early steps leading to lamb survival. *Reprod. Nutr. Dev.* **2006**, *46*, 431–446. [CrossRef] [PubMed]
71. Gupta, A.; Paria, A. Transition from fetus to neonate. *Surgery* **2019**. [CrossRef]
72. Hoseth, E.; Joergensen, A.; Ebbesen, F.; Moeller, M. Blood glucose levels in a population of healthy, breast fed, term infants of appropriate size for gestational age. *Arch. Dis. Child. Fetal Neonatal Ed.* **2000**, *83*, 117–119. [CrossRef]
73. Aoki, T.; Ishii, M. Hematological and biochemical profiles in peripartum mares and neonatal foals (Heavy Draft Horse). *J. Equine Vet. Sci.* **2012**, *32*, 170–176. [CrossRef]
74. González-Lozano, M.; Trujillo-Ortega, M.E.; Becerril-Herrera, M.; Alonso-Spilsbury, M.; Rosales-Torres, A.M.; Mota-Rojas, D. Uterine activity and fetal electronic monitoring in parturient sows treated with vetrabutin chlorhydrate. *J. Vet. Pharmacol. Ther.* **2010**. [CrossRef] [PubMed]
75. Mota-Rojas, D.; Martinez-Burnes, J.; Ortega, M.E.T.; López, A.; Rosales, A.M.; Ramirez, R.; Alonso-Spilsbury, M. Uterine and fetal asphyxia monitoring in parturient sows treated with oxytocin. *Anim. Reprod. Sci. Sci.* **2005**, *86*, 131–141. [CrossRef]
76. Herpin, P.; Le Dividich, J.; Hulin, J.C.; Fillaut, M.; De Marco, F.; Bertin, R. Effects of the level of asphyxia during delivery on viability at birth and early postnatal vitality of newborn pigs. *J. Anim. Sci.* **1996**, *74*, 2067–2075. [CrossRef]
77. Marchant, J.N.; Rudd, A.R.; Mendl, M.T.; Broom, D.M.; Meredith, M.J.; Corning, S.; Simmins, P.H. Timing and causes of piglet mortality in alternative and conventional farrowing systems. *Vet. Rec.* **2000**, *147*, 209–214. [CrossRef]
78. Lawler, D.F. Neonatal and pediatric care of the puppy and kitten. *Theriogenology* **2008**. [CrossRef]
79. Mota-Rojas, D.; Rosales, A.M.; Trujillo, M.E.; Orozco, H.; Ramírez, R.; Alonso-Spilsbury, M. The effects of vetrabutin chlorhydrate and oxytocin on stillbirth rate and asphyxia in swine. *Theriogenology* **2005**, *64*, 1889–1897. [CrossRef]
80. Mota-Rojas, D.; Nava-Ocampo, A.A.; Trujillo, M.E.; Velázquez-Armenta, Y.; Ramírez-Necoechea, R.; Martínez-Burnes, J.; Alonso-Spilsbury, M. Dose minimization study of oxytocin in early labor in sows: Uterine activity and fetal outcome. *Reprod. Toxicol.* **2005**, *20*, 255–259. [CrossRef]
81. MacDonald, H.M.; Mulligan, J.C.; Allen, A.C.; Taylor, P.M. Neonatal asphyxia. I. Relationship of obstetric and neonatal complications to neonatal mortality in 38,405 consecutive deliveries. *J. Pediatr.* **1980**, *96*, 898–902. [CrossRef]
82. Vaala, W.E. Peripartum asphyxia syndrome in foals. *AAEP Proc.* **1999**, *45*, 247–253.
83. González-Lozano, M.; Mota-Rojas, D.; Orihuela, A.; Martínez-Burnes, J.; Di Francia, A.; Braghieri, A.; Berdugo-Gutiérrez, J.; Mora-Medina, P.; Ramírez-Necoechea, R.; Napolitano, F. Review: Behavioral, physiological, and reproductive performance of buffalo cows during eutocic and dystocic parturitions. *Appl. Anim. Sci.* **2020**, *36*, 407–422. [CrossRef]

84. Mota-Rojas, D.; Martínez-Burnes, J.; Napolitano, F.; Domínguez-Muñoz, M.; Guerrero-Legarreta, I.; Mora-Medina, I.; Ramírez-Necoechea, R.; Lezama-García, K.; González-Lozano, M. Dystocia: Factors affecting parturition in domestic animals. *CAB Rev. Perspect. Agric. Vet. Sci. Nutr. Nat. Resour.* **2020**, *15*. [CrossRef]
85. Mota-Rojas, D.; De Rosa, G.; Mora-Medina, P.; Braghieri, A.; Guerrero-Legarreta, I.; Napolitano, F. Dairy buffalo behaviour and welfare from calving to milking. *CAB Rev. Perspect. Agric. Vet. Sci. Nutr. Nat. Resour.* **2019**. [CrossRef]
86. Zamora-Moran, E. Metabolic blood gas disorders. In *Acid-Base and Electrolyte Handbook for Veterinary Technicians 2017*; John Wiley and Sons: Chichester, UK, 2017; pp. 121–135.
87. Ross, M.G.; Gala, R. Use of umbilical artery base excess: Algorithm for the timing of hypoxic injury. *Am. J. Obstet. Gynecol.* **2002**, *187*, 1–9. [CrossRef]
88. Siristatidis, C.; Salamalekis, E.; Kassanos, D.; Loghis, C.; Creatsas, G. Evaluation of fetal intrapartum hypoxia by middle cerebral and umbilical artery Doppler velocimetry with simultaneous cardiotocography and pulse oximetry. *Arch. Gynecol. Obstet.* **2004**, *270*, 265–270. [CrossRef]
89. Manning, M. Electrolyte disorders. *Vet. Clin. N. Am. Small Anim. Pract.* **2001**, *17*, 8–13. [CrossRef]
90. Best, P.R. Anaesthesia during pregnancy and reproductive surgery. *Anesthesiology* **1988**, *69*, 475–510.
91. Mota-Rojas, D.; Orihuela, A.; Strappini-Asteggiano, A.; Cajiao, P.M.N.; Aguera-Buendia, E.; Mora-Medina, P.; Ghezzi, M.; Alonso-Spilsbury, M. Teaching animal welfare in veterinary schools in Latin America. *Int. J. Vet. Sci. Med.* **2018**, *6*, 131–140. [CrossRef] [PubMed]
92. Villanueva-García, D.; Mota-Rojas, D.; Martínez-Burnes, J.; Olmos-Hernández, A.; Mora-Medina, P.; Salmerón, C.; Gómez, J.; Boscato, L.; Gutiérrez-Pérez, O.; Cruz, V.; et al. Hypothermia in newly born piglets: Mechanisms of thermoregulation and pathophysiology of death. *J. Anim. Behav. Biometeorol.* **2021**, *9*, 2102. [CrossRef]
93. Mota-Rojas, D.; Titto, C.G.; Orihuela, A.; Martínez-Burnes, J.; Gómez-Prado, J.; Torres-Bernal, F.; Flores-Padilla, K.; Carvajal-de la Fuente, V.; Wang, D. Physiological and behavioral mechanisms of thermoregulation in mammals. *Animals* **2021**, *11*, 1733. [CrossRef] [PubMed]
94. Orihuela, A.; Mota-Rojas, D.; Strappini, A.; Serrapica, F.; Braghieri, A.; Mora-Medina, P.; Napolitano, F. Neurophysiological mechanisms of mother–young bonding in buffalo and other farm animals. *Animals* **2021**, *11*, 1968. [CrossRef] [PubMed]
95. Mota-Rojas, D.; Pereira, A.; Wang, D.; Martínez-Burnes, J.; Ghezzi, M.; Lendez, P.; Bertoni, A.; Geraldo, A.M. Clinical applications and factors involved in validating thermal windows used in infrared thermography to assess health and productivity. *Animals* **2021**, *11*, 2247. [CrossRef]

Article

Neurohormonal Profiles of Assistance Dogs Compared to Pet Dogs: What Is the Impact of Different Lifestyles?

Manuel Mengoli [1,2], Jessica L. Oliva [1], Tiago Mendonça [1,2], Camille Chabaud [1], Sana Arroub [1], Céline Lafont-Lecuelle [1], Alessandro Cozzi [1], Patrick Pageat [1,2] and Cécile Bienboire-Frosini [1,*]

[1] Research Institute in Semiochemistry and Applied Ethology (IRSEA), 84400 Apt, France; manuelmengoli@mac.com (M.M.); jessicaleeoliva@gmail.com (J.L.O.); t.mendonca@group-irsea.com (T.M.); c.chabaud@group-irsea.com (C.C.); s.arroub@group-irsea.com (S.A.); celine.lecuelle@sfr.fr (C.L.-L.); a.cozzi@group-irsea.com (A.C.); p.pageat@group-irsea.com (P.P.)
[2] Clinical Ethology and Animal Welfare Centre (CECBA), 84400 Apt, France
* Correspondence: c.frosini@group-irsea.com; Tel.: +33-490-750-618

Citation: Mengoli, M.; Oliva, J.L.; Mendonça, T.; Chabaud, C.; Arroub, S.; Lafont-Lecuelle, C.; Cozzi, A.; Pageat, P.; Bienboire-Frosini, C. Neurohormonal Profiles of Assistance Dogs Compared to Pet Dogs: What Is the Impact of Different Lifestyles? *Animals* **2021**, *11*, 2594. https://doi.org/10.3390/ani11092594

Academic Editors: Angelo Gazzano and Asahi Ogi

Received: 30 June 2021
Accepted: 26 August 2021
Published: 3 September 2021

Simple Summary: Dogs are currently involved in various roles in our society beyond companionship. The tasks humans assign to them impact their daily life and can sometimes create stressful situations, possibly jeopardizing their welfare. For example, assistance dogs need to manage their emotions in various challenging situations and environments. Thus, the capacity to cope with emotional stress is highly desirable in assistance dogs (~40% of assistance dogs fail to complete their education program). The emotional and stress responses are guided by brain processes involving neuromodulators. Neurohormonal profiling of these dogs can: (i) give cues about their emotional suitability to fulfill an assistance role; (ii) enhance their selection; and (iii) help to assess and improve their welfare state during the training course. We compared basal blood levels of three neuromodulators of interest between two populations, assistance vs. pet dogs. We found significantly different concentrations of oxytocin, a neuromodulator involved in social behavior. Levels of prolactin, a putative marker of chronic stress, were higher (although not statistically significant) and variable in assistance dogs. Dogs' age also seemed to influence the various neuromodulators levels. These findings highlight the impact of different lifestyles undergone by dogs and the possibility to use neurohormonal profiling to monitor their effect on the dogs' welfare and stress state.

Abstract: Assistance dogs must manage stress efficiently because they are involved in challenging tasks. Their welfare is currently a fundamental issue. This preliminary study aimed to compare assistance dogs (AD; n = 22) with pet dogs (PD; n = 24), using blood neuromodulator indicators to help find biomarkers that can improve the AD breeding, selection, training, and welfare monitoring. Both populations originated from different breeds, are of different ages, and had different lifestyles. Basal peripheral concentrations of prolactin (PRL), serotonin (5-HT), free (fOT) and total (tOT) oxytocin were measured by immunoassays. Multiple linear regressions were performed to assess the effect of activity, age, sex, and their interactions on these parameters. Correlations between neurohormonal levels were analyzed. No interactions were significant. fOT and tOT concentrations were significantly influenced by age ($p < 0.0001$ and $p = 0.0002$, respectively) and dogs' activity ($p = 0.0006$ and $p = 0.0277$, respectively). A tendency was observed for age effect on PRL ($p = 0.0625$) and 5-HT ($p = 0.0548$), as well as for sex effect on tOT ($p = 0.0588$). PRL concentrations were heterogenous among AD. fOT and tOT were significantly but weakly correlated (Pearson's r = 0.34; $p = 0.04$). Blood prolactin, serotonin, and oxytocin may represent biomarkers to assess workload and chronic stress-related responses in ADs and eventually improve their selection and training.

Keywords: *Canis lupus familiaris*; guide dogs; biomarker; neuromodulator; serotonin; free oxytocin; total oxytocin; prolactin; stress

1. Introduction

"Assistance dogs" (AD) include (i) service dogs; i.e., dogs that assist humans with mobility impairments as well as medical detection dogs and dogs assisting people with psychiatric impairments; (ii) hearing dogs; i.e., dogs that assist humans that have hearing impairments and (iii) guide dogs; i.e., dogs that assist humans that have visual impairments [1]. These dogs are often involved in complicated tasks (i.e., emotionally and cognitively challenging) in their daily life: they need to make the right behavioral choice while facing ambivalent situations, thereby adequately balancing their cognitive–emotional responses [2]. The term "ambivalent" refers to the notion of choice that can be equivocal. The assistance dogs must often choose between two situations, and sometimes go against their owner's indications if necessary (e.g., in case of danger). For instance, the dog needs to show to not be over-reactive during walking, with good capabilities in filtering the external stimulation (social or environmental), being comfortable with and giving safety to the handler [3,4]. Good emotional balance and capacity to cope with emotional stress are therefore the most desired characteristics in trained guide dogs [5]. This is why dogs are often selected to be as calm as possible ("stability"/"passivity"), displaying no sign of aggression to other animals and humans ("docility") and with a predisposition to demonstrate a good behavioral adaptation to their environment ("adaptability"). Conversely, distraction, shyness, and fearfulness are undesirable behavioral traits [6–8]. Breeds are often selected for assistance work on the basis of the degree of expression of these characteristics [9]. Individual characteristics will also influence the breeding and the educational protocols, with many dogs that fail, then are rejected or reoriented (to other assistance activities or offered for adoption) [7,10]: for instance, currently, the French guide dogs network face an approximate 40 percent loss of dogs due to physical or behavioral problems [11].

Only several research studies have investigated factors associated with long-term success in AD [2,7,8]. These dogs need to balance their emotional responses and try to find the best coping strategies to reduce their stress-related responses in a daily working routine. These emotional and stress responses are guided by brain processes involving neuromodulators. Investigating these neuromodulators in AD is an interesting approach to better understand their emotional and stress responses. Additionally, they can serve as indicators to help to assess the dogs' individual emotional characteristics, hence their future chances of success in the AD programs. Among the neuromodulators which have an interest in the context of AD evaluation and follow-up, we focused on prolactin, serotonin, and oxytocin.

Prolactin (PRL) is a polypeptide neurohormone that plays multiple homeostatic roles in the organism in conjunction with the dopaminergic system [12]. PRL blood levels increase during acute or chronic stress (due to psychological or physiological stressors) and increased levels of PRL have been linked to emotional disorders and anxiety-related behaviors in several mammalian species [13–15]. PRL may change in response to psychological stress also in humans, where different studies have shown a positive correlation between anxiety and negative emotions and serum PRL [16–18]. Previous studies have indicated the interest in evaluating serum PRL in anxious dogs [19]. On the contrary, recent studies [20] highlighted that in sheltered castrated male dogs' PRL concentrations correlate neither to stress scores nor to fear behaviors, and that a weak negative correlation between cortisol and prolactin levels was possible. In previous research, we found that AD presented higher levels of blood PRL than a pet dog population in a controlled situation [11,21], emphasizing the interest of further investigating this neuromodulator.

Serotonin or 5-hydroxytryptamine (5-HT) is a monoamine found in the central nervous system, gastrointestinal tract, and blood. It has broad physiological functions as a neuromodulator [22]. Serotonergic neurons have numerous projections throughout the brain, provide 5-HT to the rest of the central nervous system, and can modulate limbic-HPA activity [23]. 5-HT was demonstrated to be an interesting neuromodulator to study dog behavioral and emotional responses since low levels of 5-HT were found to be associated with animal distress and behavioral disorders, such as impulsiveness, loss of motor control, and aggression in certain dogs [24,25]. It was shown that aggressive dogs had significantly lower serum concentrations of 5-HT than non-aggressive dogs [24,26]. For AD, intraspecific and interspecific aggression are unacceptable behaviors, that are unequivocally associated with dogs' rejections, with the animal exiting the educational program and the school [6,11]. Therefore, we found it crucial to assess this neuromodulator in AD as a part of the dog's emotional evaluation before and during their educational training.

Oxytocin (OT) is a nonapeptide produced by the hypothalamus and released from magnocellular neurons that project to the posterior lobe of the pituitary where it is naturally secreted into the bloodstream and acts as a hormone, mainly involved in parturition and lactation [27]. OT is also transported to various brain regions via axonal pathways where it acts as a neuromodulator, with a crucial role in many social behaviors, ranging from social memory and affiliation (sexual and parental behaviors, attachment, pair-bonding) to aggression ("mate-guarding," maternal aggression) (for reviews, see [28–31]). In particular, OT has been described as being involved in positive social intra-species interactions [32–35] and inter-species interactions, such as human-animal interactions [36–40]. In dog applied ethology, OT is often related to research on how a social interspecific interaction with humans [41,42] or social stress [43] will modify OT levels, or on the effect of OT external administration on dogs' cognition and communication and other behaviors or aptitudes [21,44–48]. For instance, intranasally applied oxytocin has been shown to increase positive expectations [44], increase the use of human social cues such as pointing [48], modulate the dogs' aggressive response to the threatening cues of a human [49], increase affiliation toward owners and conspecifics [38], increase intraspecific play behaviors [34], and influence the visual contact with the owners [45,46]. AD need to show a predisposition for positive prosocial interaction with humans, good mental flexibility, and being smartness, with an important role of OT in all these aptitudes [41,46,50].

Currently, OT has become a renowned neuromodulator, an indicator commonly investigated in behavioral and neuroendocrinological research, in particular, human–dog interactions [42,51]. Several studies have reported a coordinated release of central and peripheral OT [52,53] and showed that peripheral levels can provide a minimally invasive indicator of central state [54], from a practical point of view, when central parameters are not available. In dogs, OT concentrations have been peripherally quantified in plasma, urine, and saliva [55–57] despite important methodological considerations and caveats. OT is difficult to assay and methods to measure endogenous OT greatly vary [55,58]. This results in a lack of correlation between them and in discrepant results among the literature, hence these methods are still under development [59]. MacLean et al. [59] proposed that OT molecules may exist under different molecular conformational states, such that these different states may be differently measured by the various used methods, leading to the observed discrepancies. Among the putative molecular states of OT, Brandtzaeg et al. [60] developed a method assaying the protein-bound OT in plasma. OT is circulating in two forms, the most known free OT (fOT), and the protein-bound form, both fractions constituting the total amount of OT (tOT) in plasma. Different authors are working to describe different methods to measure total OT levels and protein-bound levels and to find what their biological functions may be [50,61,62]. Measuring only the fOT, as currently recommended [63], may, in many cases, be a confusing factor, since the fOT concentration can change because of dynamic intrinsic or extrinsic factors or substances that displace OT from proteins [58,64]. Furthermore, Uvnäs Moberg et al. [65] recently hypothesized that OT may be a principal hormone and exert its effects through its active fragments, adding more consideration to the possible importance of the different OT molecular forms.

The aims of this study were (i) to describe the neurohormonal profiles of two different populations of dogs (assistance dogs: AD and pet dogs: PD) displaying different daily routines and lifestyles in relation to their breed selection, by measuring their basal peripheral neuromodulator levels that may be of importance regarding the monitoring of their socio-emotional behaviors i.e., PRL, 5-HT, and OT; (ii) to further investigate the biological significance of the different molecular states of oxytocin by exploring both total and free oxytocin (tOT and fOT) basal levels in these two different populations of dogs; (iii) to look at the putative associations between these neuromodulators peripheral measures, considering their physiological interactions within the central nervous system [66,67].

We expect that the different lifestyles experienced by the AD and PD as well as the AD individual and breed selection may result in substantial differences in their basal levels of these circulating neurohormones involved in social behavior regulation. Beyond that, we thought establishing the neurohormonal profile of AD may help assess their future chance of success such that AD selection before entering the educational program may be refined. That is particularly important considering the percent of failures and the cost of this loss [11,21]. In addition, using neurohormonal profiling to evaluate their emotional responses to the workload and challenging situations they face during their training may help assess their welfare. This may conduct to an improvement of the educational programs and hopefully, in the longer term, to establish guidelines about the training of AD.

2. Materials and Methods

2.1. Subjects

The dog population involved was the same previously presented in Oliva et al. [21]: fifty-one dogs were involved at the beginning of the trial and checked by a veterinary behaviorist to look for possible medical or behavioral reasons to exclude them. Five dogs were excluded because 1 blood drawing did not provide a sufficient sample's volume, 1 was discovered pregnant, 2 showed a phobic response to humans and possible aggression, and 1 had moved during the trials. The final sample, therefore, included 46 dogs that were blood sampled, considering 22 young AD (9 Males and 13 Females) with a mean age of 1.10 ± 0.17 years and 24 PD (16 Males and 8 Females) with a mean age of 4.98 ± 3.13 years. The dogs did not present unhealthy conditions, and they were not under any medication or showing any behavioral or emotional disorders (as shown by the use of the validated emotional disorders evaluation scale (EDED), see [21]). When females were entire, they were tested at least 2 months out of estrus.

All AD came from the Frédéric Gaillanne Foundation (FGF; L'Isle-sur-la-Sorgue, France), an AD training school where they were in training to be guide dogs or dogs involved in assistance of children with autism spectrum disorder. All dogs recruited from the FGF were in training and still living part time with the same human caregiver who had been assisting the FGF to raise them since they were between 8 and 10 weeks old ("foster families"). FGF dogs entered the school at about 12 months, and they were involved in a specific training for at least one and half years, from Monday to Friday at the school, excluding the weekends and the holidays, when they were with their foster families. They were following the same diet and performed similar exercise and daily activities, proportionally to their age. All dogs came from the same breeding selection: they were Labrador Retriever or Saint-Pierre Dogs (a crossed breed of Bernese Mountain Dogs and Labrador Retriever) (Table 1).

PD were a heterogeneous clinical population. All animals were older than 10 months, of different breeds (Table 1), sizes, following different diets, and with different daily routines. They were recruited from owners that heard about the trial due to publicity on social media and social networks; other animals were referred by other clinics. Hence, the two dog populations of this study underwent a different daily routine in terms of environmental stimulation, richness of intra- and interspecific social interactions, and mental and physical training.

2.2. Ethical Statement

This study was performed in accordance with French (2013-118) and European law (2010/63/EU) on the protection of animals used for scientific purposes. The study was approved by the IRSEA Ethics Committee C2EA125 (approval number AFCE_201605_02).

Table 1. Breed differences between assistance dogs (AD) and pet dogs (PD).

Origin of Dogs			
Pet Dogs (PD)		**Assistance Dogs (AD)**	
Breed	**N**	**Breed**	**N**
Mixed	3	Bernese Mountain dog × Labrador (St Pierre)	15
Border Collie	3	Labrador	6
Welsh Corgi Cardigan	2	Labrador × golden retriever	1
Australian Shepard	2		
German Shepard	2		
Labrador	2		
Border Collie Cross	1		
Australian Shepard × German Shepard	1		
Bernese Mountain Dog	1		
Labrador × Boxer	1		
Malinois	1		
Samoyed	1		
Poodle	1		
Yorkshire Terrier	1		
Westie	1		
Pinscher × Chihuahua	1		

Adapted from Oliva et al. [21].

2.3. Blood Sampling and Processing

All dogs were accommodated in the same consultation room of the CECBA (Clinical Ethology and Animal Welfare Centre) facilities (Apt, France), with the same environmental temperature (21–23 °C) at the same time schedule. Moreover, all owners were asked to keep their dogs indoors the night before the blood drawing to minimize the effect of ambient temperature on neurohormones' secretion.

With the help of an operator, blood was taken by a veterinarian from the cephalic vein or the jugular vein in the least stressful conditions possible and only if the immobilization was feasible. An amount of 8–10 mL of total blood per dog was collected into pre-chilled pink EDTA-Aprotinin vacuum tubes (BD® tubes, Elvetec, Pusignan, France) to collect plasma and approximately 1 mL was collected into red vacuum tubes with gel separator (Vacuette® Greiner Bio-One, Alcyon, Paris, France) to collect serum. The pink tubes were immediately transferred to an ice compartment where they remained until centrifugation, while red tubes remained at room temperature for between 30- and 180-min. Samples were centrifuged at 1600× *g* for 15 min at 4 °C. Serum and plasma were transferred to a plastic tube and stored at −20 °C until analyses. Because of the necessity to differently concentrate and process the plasma/serum for the various analyses, certain samples were of insufficient volume to be assayed for each parameter. Then, the number of final samples immunoassayed and used for the statistical protocols was different regarding the neuromodulators (PRL: 21 AD and 24 PD; 5-HT: 22 AD and 24 PD; fOT: 20 AD and 21 PD; tOT: 21 AD and 19 PD).

2.4. Neurohormones Analyses

Neurohormone analyses were performed by the IRSEA labs (Research Institute in Semiochemistry and Applied Ethology, Apt, France).

PRL concentrations from canine serum were measured using the canine Prolactin ELISA kit (#DEV9944, Demeditec Diagnostics®, Kiel, Germany), according to the manufacturer's instructions. The lower detection limit for canine prolactin was 0.4 ng/mL (assay range: 0–80 ng/mL).

Serum 5-HT levels were measured by enzyme immunoassay (Serotonin ELISA kit #ADI-900-175, EnzoLifeSciences, Villeurbanne, France) according to the manufacturer's instructions and the validated method detailed in the referred application note [68]. This kit had already been used in previous studies assaying 5-HT in a variety of animal species, including the dog [69]. The kit's sensitivity is 0.293 ng/mL with an assay range between 0.49 and 500 ng/mL.

Plasma fOT was assayed using the Oxytocin ELISA kit from Cayman Chemical (#500440, Ann Arbor, MA, USA) according to the manufacturer recommendations, including the plasma solid-phase extraction on C18 columns (Hypersep 1 g, Thermo Fisher Scientific, Illkirch, France), allowing a two-times concentration factor, as explained in [21]. Plasma tOT was assayed using the same ELISA kit after reduction/alkylation and protein precipitation (R/A PPT) procedures, which release bound OT from plasma proteins and allow for the detection of much higher concentrations of OT in dog plasma, according to the previously published methods [60,61]. Chemicals used in this procedure (acetonitrile, dithiothreitol, iodoacetamide) were provided by Sigma-Aldrich (Saint-Quentin-Fallavier, France).

2.5. Statistical Analysis

All data were analyzed in SAS 9.4 software (Copyright (©) 2002–2012 by SAS Institute Inc., Cary, NC, USA). The significance threshold was classically fixed at 5% for all the following statistical analyses. First, preliminary statistical tests were performed to evaluate the effect of the dog's activity (PD vs. AD) on age and sex and determinate if these last should be included in the statistical analysis. For the age variable, comparison between the two groups (AD and PD) was performed with the Wilcoxon/Mann–Whitney test with the NPAR1WAY procedure. Conditions of normality and heteroscedasticity were tested respectively with the UNIVARIATE and TTEST procedures and have not been checked. For sex and activity effects, their association was studied with the Chi-Square test with the FREQ procedure.

Then, for PRL, 5-HT, fOT, and tOT, a multiple linear regression was performed using the GLM procedure. Effects of activity (AD and PD), sex (male and female), age, and the three corresponding interactions were tested in the complete model. Conditions of normality of residues and homoscedasticity were tested using respectively the UNIVARIATE and the GPLOT procedures. For the age effect (quantitative variable), the linearity assumption was verified with the GPLOT procedure. For fOT and tOT, these conditions were verified. For PRL and 5-HT, normality of residues was not verified; a Box-Cox transformation was applied to data using the TRANSREG procedure, which allowed to verify the normality and then the homoscedasticity. The three interactions were removed of the model because they were not significant.

The correlation between fOT and tOT in both dog populations was assessed using Pearson correlation analyses with the CORR procedure since both sets of data were normal. The correlations between 5-HT, fOT and PRL were evaluated using Spearman correlation analyses due to the CORR procedure since the normality of data was not verified.

3. Results

Blood concentrations of PRL, 5-HT, fOT, and tOT in dogs are shown in Table 2.

Table 2. Descriptive data for basal peripheral concentrations levels of prolactin (PRL), serotonin (5-HT), free and total oxytocin (fOT and tOT) for assistance dogs (AD) and pet dogs (PD).

Dogs	Neurohormones	Mean	SD	Median	Min	Max	Lower Quartile	Upper Quartile
AD	PRL (ng/mL)	14.2	18.6	6.8	0.2	71.3	2.5	18.7
	5-HT (ng/mL)	1220.3	615.9	1114.5	144.6	2664.7	876.6	1339.4
	fOT (pg/mL)	27.1	9.7	26.0	12.7	47.2	19.1	35.1
	tOT (pg/mL)	1136.7	320.7	1113.6	517.1	1774.1	1018.7	1362.1
PD	PRL (ng/mL)	6.3	8.6	3.2	0.2	38.8	0.7	10.0
	5-HT (ng/mL)	868.5	450.1	702.1	359.2	1958.3	525.8	1064.5
	fOT (pg/mL)	30.8	17.2	27.8	5.4	66.5	15.8	39.9
	tOT (pg/mL)	1032.4	405.4	1206.3	347.7	1535.6	635.6	1371.6

SD: standard deviation.

3.1. Preliminary Statistical Analyses to Assess the Similarity between AD and PD Groups Regarding the Age and Sex Factors

A significant difference was observed between the two groups (AD and PD) for age (Wilcoxon/Mann–Whitney, N = 46, Z = −5.0057, $p < 0.0001$) (AD: mean ± SD = 1.10 ± 0.17 years versus PD: mean ± SD = 4.98 ± 3.13 years). No significant difference is observed between the two groups for sex (Chi-Square, N = 46, DF = 1, $\chi^2 = 3.0693$, $p = 0.0798$) (female: 13 AD versus 8 PD and male: 9 AD versus 16 PD).

3.2. Multiple Linear Regressions

Only the age factor appeared to be significantly different between the AD and PD group. However, we observed a tendency for the sex factor. Then, we decided to perform a multiple linear regression analysis including the three factors: activity (AD vs. PD), sex (M vs. F), age (as a quantitative variable) for each of the four parameters (PRL, 5-HT, fOT, tOT). We also investigated the interactions between these three factors (Table 3). None was significant in the complete model. Then, we could remove them from the complete model to focus on each of the three factors independently (simplified model). Figure 1 displays graphical representations of the linear models obtained for each parameter according to the three factors tested (Age, Sex, and Activity).

Table 3. Results of the multiple linear regressions for PRL, 5-HT, fOT and tOT variables including activity, age, and sex effects and their interactions.

Variable	Complete Model			Simplified Model		
	Interaction	F Value	*p*-Value	Factor	F Value	*p*-Value
PRL	Activity × Sex	0.20	0.6566	Activity	0.02	0.9014
	Age × Activity	1.00	0.3245	Age	3.67	0.0625
	Sex x Age	0.71	0.4045	Sex	0.53	0.4707
5-HT	Activity × Sex	0.40	0.5330	Activity	0.10	0.7518
	Age × Activity	0.00	0.9799	Age	3.90	0.0548
	Sex × Age	1.95	0.1703	Sex	0.01	0.9058
fOT	Activity × Sex	1.33	0.2564	Activity	13.96	0.0006
	Age × Activity	2.17	0.1497	Age	19.27	<0.0001
	Sex × Age	0.49	0.4896	Sex	0.05	0.8160
tOT	Activity × Sex	0.18	0.6776	Activity	5.27	0.0277
	Age × Activity	1.93	0.1737	Age	16.85	0.0002
	Sex × Age	2.01	0.1658	Sex	3.81	0.0588

In all panels, we clearly observed the linear relationship between the neurohormonal parameters and age for each activity and sex group. We could also remark the greater dispersion of points around the regression line for PD due the inequal repartition of dogs according to age, with AD being significantly younger (less than 2 years old), as previously stated.

Detailed results of the significant statistical models are shown in Table 4.

Activity, age, or sex had no significant effect on PRL levels that is shown by the close regression lines for sex and activity observed in Figure 1a. Only a tendency was observed for the factor age (Table 3), with older dogs tending to display lower levels of PRL ($\beta = -0.2570$; SE = 0.1342), as also displayed by the negative weak slopes of the regression lines in Figure 1a. Nevertheless, looking at the descriptive data (Table 2), AD displayed mean and median serum PRL levels at least two times higher than PD. We noticed an important heterogeneity of serum PRL concentrations among the AD compared to PD, with a standard deviation (SD) of 18.6 vs. 8.6 ng/mL, respectively, and a larger range of concentrations in AD than in PD, due to the different maximum values reached in each population: 71.3 ng/mL in AD vs. 38.8 ng/mL in PD. The Fisher test assessing the equality of variances gave the following outcomes: F-value = 4.73, p = 0.0005, confirming that variances are heterogenous within both populations.

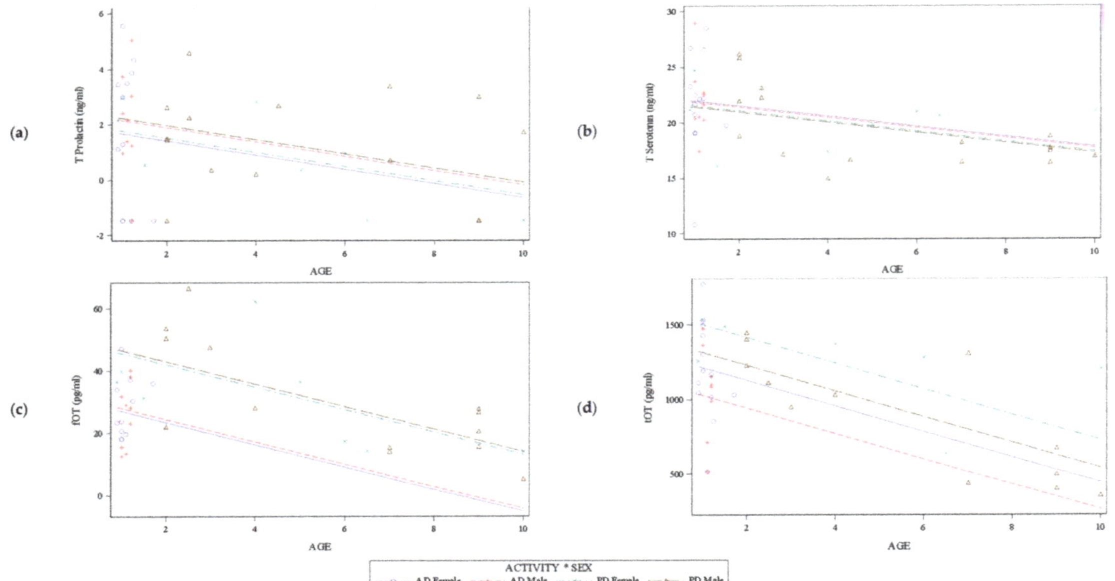

Figure 1. Linear regression models of T (Transformed) prolactin (**a**), T (transformed) serotonin (**b**), free oxytocin (fOT) (**c**), and total (tOT) (**d**) according to age, activity, and sex. Individual values are marked with points of different shapes and the regression lines are shown in solid or broken lines of different colors according to the legend. The parallel regression lines highlight that there is no interaction between the factors. Merged or closed regression lines indicate that there is no effect of the factors' activity/sex, or not much difference between them, respectively. Identical slopes of the regression lines denote similar effects of the age, regardless of the dogs' activity/sex.

Table 4. Unstandardized regression coefficients for each predictor having significant effects following the multiple linear regression analyses for fOT and tOT variables.

Variable	Predictor	β	SE	p-Value
fOT	Activity (Reference PD)	−18.5189	4.9556	0.0006
	Age	−3.5911	0.8180	<0.0001
tOT	Activity (Reference PD)	−285.8182	124.5404	0.0277
	Age	−86.5206	21.0776	0.0002

β: unstandardized regression coefficients or estimates; SE: standard error; PD: pet dogs.

Similarly, there was no significant effect of the activity, age, or sex on 5-HT peripheral concentrations, as showed by the merged regression lines in Figure 1b. However, a substantial tendency appeared for the factor age (Table 3): the older the dogs were, the lower the 5-HT levels tended to be (β = −0.4573; SE = 0.2315), also revealed by the negative weak slopes of the regression lines in Figure 1b.

Both factors activity and age had significant effects on the fOT concentrations. Age appeared to be a confounding factor since the older the dogs, the lower the fOT ($p < 0.0001$) (Table 4). Thus, for each additional year of age of the dog, the fOT level decreased by an average of 3.59 pg/mL, regardless of the activity of the dog. After adjustment for age, fOT levels were on average 18.52 pg/mL lower in AD than in PD (p = 0.0006). This effect of activity was also denoted by the distance between the regression lines of each activity group in Figure 1c. Conversely, the regression lines of both sexes for the same activity group were merged, indicating no sex effect, in accordance with the insignificant *p values* of the simplified model shown in Table 3.

Similarly, both activity and age had significant effects on tOT concentrations and the age appeared to be a confounding factor too: the older the dogs, the lower the tOT (p = 0.0002) (Table 4). Thus, for each additional year of age of the dog, the tOT level decreased by an average of 86.52 pg/mL, regardless of the activity of the dog. After adjustment for age, tOT levels were on average 285.82 pg/mL lower in AD than in PD (p = 0.0277). Regarding the sex effect (Table 3), there was also a tendency on tOT concentrations: females tended to display higher tOT levels than males (β = 187.1804; SE = 95.9044). These effects are also suggested by the distance between the four regression lines observed in Figure 1d.

There was a significant correlation between fOT and tOT levels in the entire dog population, which was positive but weak (N = 37; Pearson's r = 0.34; p = 0.04). The correlation between 5-HT and PRL was not significant (N = 45; Spearman's Rho = 0.157; p = 0.30). The correlation between 5-HT and fOT showed a tendency and was positive and weak (N = 41; Spearman's Rho = 0.290; p = 0.07). Finally, the correlation between fOT and PRL was not significant (N = 41; Spearman's Rho = −0.014; p = 0.93).

4. Discussion

In this study, we aimed to describe basal blood levels of certain neuromodulators involved in socio-emotional behaviors i.e., PRL, 5-HT, and OT (free and total) in AD vs. PD (with AD having different ages, different daily routines and being of different breeds) and to explore their putative relationships. We found that fOT and tOT concentrations were significantly influenced by the dogs' age and activity (as PD or AD): younger dogs had higher levels of fOT and tOT, whereas ADs had lower levels of fOT and tOT. As the AD were significantly younger than PD, both age and activity effects were confounded, thus explaining why the simple observation of descriptive data did not allow to show these outcomes. We also found tendencies on the effect of sex on tOT concentrations (higher in female) and the effect of age on PRL and 5-HT concentrations, following the same direction: the younger the dogs are, the higher their levels of peripheral neurohormones are. We also observed that peripheral PRL levels are heterogeneous in the population of AD compared to PD, which is the expression of a greater variability of this parameter within AD population. Additionally, this study showed that fOT and tOT plasma measures are significantly but weakly correlated, with an association of only 11.6%. Other neurohormonal parameters were not significantly correlated between them.

4.1. Prolactin

Median serum PRL levels found in AD and PD can be considered as normal according to the reference values provided by the ELISA manufacturer user's guide and matched those described in the literature for normal dogs [70,71]. In this study, there were no significant differences in serum PRL between AD and PD, although mean and median PRL concentrations were at least 2-times higher in AD than in PD. A precedent study [21] with the same population involved in a different protocol (sub-divided in three groups receiving different treatments and undergoing several blood samplings) showed the average prolactinemia was significantly higher in AD vs. PD, while here, we only considered the basal level of serum prolactin to reflect the basal state of these two populations. Importantly, the age effect was not considered in [21]. We found here that the age tended to influence the blood PRL levels with younger dogs displaying higher PRL basal concentrations. This new statistical approach on the same dog population could in fact explain the previous results as well as the descriptive data we observed here on AD: the higher PRL levels might be due to the younger age of AD rather than their activity. Further studies comparing AD and PD of the same age can help to clarify this question. Alternatively, the higher prolactinemia, we also observed in descriptive data of AD, could be related to breed/genetic characteristics, as suggested by previous literature [72,73]. AD are notably selected on desirable behavioral characteristics (meaning that only animals which do not show aggressiveness, chronic phobias, or anxiety are kept for reproduction purposes), which can originate from the presence of genetic elements involved in the neuromodulation of behaviors/emotional responses. In turn, these genetic characteristics can be associated with the dog's breeds, which are more commonly selected for AD recruitment.

This higher prolactinemia observed from descriptive data may also be linked to a chronic-stress response: serum PRL concentrations have already been associated with emotional disorders in dog [19]. Because of the inhibitory control of dopamine on PRL secretion [74], PRL represents an interesting biomarker in behavioral medicine that may reflect the dopaminergic system activity, previously described as correlated with different anxious states in dogs [19,75]. Usefully, the levels of blood PRL are more stable than those of dopamine, strongly influenced by its more variable clearance features [76], thus making PRL easier to analyze. Thus, investigating the dopaminergic system activity via the measurement of blood PRL can help assess the chronic-stress responses and anxiety in dogs and follow-up their welfare states. This is of particular interest in AD who must face substantial workload and challenging situations in their daily life. However, the exact influence of age and breed on PRL levels should be determined before using it as a reliable biomarker of stress and emotional responses.

The important heterogeneity of serum PRL concentrations among the AD compared to PD is also noteworthy. PD were dogs come from different litters with no possibility of genetic familiarity, with different ages, sizes, breeds and with a different daily routine, whereas the AD population represents a population of relatives from a genetics point of view, with the same diet, size and with a similar daily routine (the same training plan and daily life at the school). Thus, PRL unexpectedly seemed to represent a more variable parameter in the AD population than in the PD population of this study. This possible individual variability within a homogenous dog population should also be considered to examine PRL as a biomarker.

4.2. Serotonin

In our study, there was no significant effect of dog's activity on 5-HT levels, whereas dogs' age tended to have an influence on it. This may partly explain the apparent difference between mean peripheral 5-HT concentrations of AD and PD that we can observe in the descriptive data (Table 2) since AD are significantly younger than PD. This may represent an interesting feature especially for young AD in training, facing workload and challenging situations since 5-HT has patience-promoting properties and is related to coping and stress moderation [77,78], implying that their welfare may be preserved in this training context [79]. However, this age-effect tendency was not found in previous studies [80–82]. Conversely, breed, diet, environment, exercise, and housing seemed to influence 5-HT concentrations in dog serum according to the literature [25,69,83]. Authors have shown considerable interbreed variation in canine serum 5-HT concentrations [25,81,82]. The population of AD we included in this study all belong to Labrador-related breeds. Therefore, breed and selection features may also fully or partly explain our descriptive data observations.

The serum 5-HT concentrations found in PD were similar to those reported by previous studies [25,69,82] but are higher than those described by others [79–81]. These differences can arise from the variable characteristics (other than the breed and the age) of the dog population used in each study: shelter dogs, obese dogs, dogs with valvular diseases, etc. They can also arise from different preanalytical factors and analytical procedures used among these studies [80]. For instance, the antibodies' affinity and specificity are crucial features in antibody-based assays, and they can differ across the available ELISA commercial kits, resulting in measurement variability. Here, we used an ELISA kit previously validated for the use in dogs' serum, following the application note of the manufacturer [68].

Despite the fact that peripheral findings may not reflect central serotonergic activity, as highlighted by Alberghina et al. [79], serum 5-HT is still considered an accessible model to study the response in the central serotonin system [84], especially in dogs considering the simplicity of the methodology [24]. Alberghina et al. [79] demonstrated a weak positive relationship between levels of serum 5-HT and sociability of dogs toward human (in terms of approaching, handling, playing). The latter is of utmost importance for AD population and their training. Behavioral characteristics such as handedness, ability to remain focused, a good emotional balance and positive prosocial behaviors are fundamental in AD and the subjects not presenting these behaviors are always reoriented or rejected [11], including eliminated from reproduction for breeding. Because of its link with coping and docility vs. aggressive behaviors, 5-HT remains an interesting neurohormonal biomarker for AD follow-up.

4.3. Free and Total Oxytocin

In our study, age significantly influenced plasma fOT and tOT levels. After adjustment for age, plasma fOT and tOT concentrations were also found to be significantly different according to dogs' activity as AD or PD, with lower levels of fOT and tOT in AD than in PD. This was unexpected since OT has often been associated with sociability, friendliness, and docility toward humans in many species, particularly in assistance animals [38,56,85]. The OT system was suggested to be involved in the dog domestication process [86,87], including if a recent study in dogs and wolves showed that life experience was more influential on OT levels than domestication, showing a positive correlation between the physical contact with the owners and the OT concentrations [88]. The AD involved in our study have a different lifestyles and life experience compared to PD; they undergo repeated separations from their foster family (that fostered them during their first year of age and then on the weekend during their training) every Monday as part of their 6–8 months-long training; from Monday to Friday, they are trained and performed different activities/exercises during the day, and at night they stay in a kennel at the guide school. This particular life experience may impact the attachment processes of AD and influence the neuromodulatory system involved in bonding, i.e., the OT system. This may explain the lower levels of fOT in AD that we found in this study. In addition, the influence of breeds (especially the particular AD breeds) on fOT and tOT basal levels is unknown and deserves further investigations.

In another population of AD, McLean et al. [42] described comparable plasma concentrations of fOT (around 19 pg/mL) before human-animal interactions. In another experiment [61], the same authors compared OT concentrations between PD and a population of AD that have been bred for affiliative and non-aggressive temperaments and found that AD had higher free and total OT, findings that were not replicated in our study. Conversely, once the age was adjusted, we found here that AD displayed lower levels of fOT and tOT. Examining the results, it appears that our population of PD displayed higher fOT concentrations than in MacLean et al.'s study, this may be due to an intrinsic population variability of PD (see [64] for more information about the effects of individual factors on the peripheral OT). Their populations of AD and PD also seemed to differ regarding the age factor (with PD and AD having a mean ages similar than in our study, approximately 4.9 years and 1.6 years respectively). Thus, it might be possible that this age difference also influenced the levels of fOT and tOT they found in the AD population compared to the PD population. We also found higher levels of tOT than in their study; despite the use of similar plasma preparation methodologies (reduction/alkylation and protein precipitation), we used different commercial assay kits (Arbor Assay #K048 vs. Cayman Chemical #500440). The variability between antibodies among OT immunoassays may explain the difference observed between the tOT levels we found and those found by MacLean et al. [61], as later hypothesized by other authors [59].

In addition, the factor sex tended to have an effect on plasma tOT concentrations in our study. This was not the case in the previously cited study of MacLean et al. [61]. However, in a previous study, we found a sex effect on tOT in horses [50], but in the opposite direction (higher tOT concentrations on males than in females). In the same study, we also investigated the relationship between fOT and tOT fractions in conjunction with the sex and lifestyles/social activities in dogs and horses, and it could differ among species [50]. In dogs, a weak correlation seems to exist between fOT and tOT, an aspect that will be interesting to further investigate to facilitate the laboratory analyses and steps of sample collection and preparation [55,59]. However, further research is needed to clarify the role of tOT, from a biological point of view, before considering possible research or clinical applications [50,60]. All these observations reinforce the interest to investigate possible neuroendocrine predictors of individual differences in social behavior and their critical roles in shaping dog social behavior, including aspects of both affiliation and aggression [42,61]. Furthermore, it is important to consider the role of OT in prosocial positive interactions and bonds with humans, be they owners, foster families, dog trainers or animal keepers [41,48,51].

4.4. Neuromodulators Correlations

Finally, no correlation between blood 5-HT, fOT (known to be the biological active form [58]) and PRL was statistically present. As such, it seems prudent to analyze the results independently. However, these results revolved around peripheral outcomes and then must be assessed for what they are: they do not dismiss central interactions between these neuromodulators. For instance, it was shown that PRL presents an important relationship with central OT that regulates PRL's secretion in many social behaviors [66]. Similarly, previous research demonstrated the link between the serotoninergic and oxytocinergic systems in the brain [67,89–91].

4.5. Implications and Perspectives

Research is ongoing to understand the causes of AD's reform or reorientation, highlighting the importance to focus on long-term factors in dogs' success [2,7,8] in order to avoid as much financial and time losses as possible because of reformed/reoriented AD, considering the high cost of each dog's education and life-care (e.g., possible total cost of 20,000–25,000 euros per dog in France [11]). It is then important to better profile behavioral tendencies and AD emotional and stress responses to increase dog welfare and to be able to detect possible distress during working activities and daily routines [92]. Monitoring basal blood levels of relevant biomarkers in AD may help to reach these objectives and eventually improve the selection of AD, as well as their breeding, by ameliorating the choice of the most "emotionally competent" individuals, as based only on their behavior. To the best of our knowledge, this study is the first to describe and profile different basal biomarkers involved in different neuromodulation systems of emotional and stress responses i.e., blood PRL, 5-HT, fOT and tOT in PD vs. AD that have different daily routines and social activities. Age was found to be a significant confounding factor for fOT and tOT. A similar tendency was observed for PRL and 5-HT. OT, which modulates bonding, social attachment, and social behavior was found to be significantly different between AD and PD, after age adjustment, with lower levels in AD that in PD. This observation may be connected to the particular life experience (repeated separations form the main attachment figure, the foster family) undergone by AD during their training.

This study also underlined the high impact of within-population variability on the other neuromodulators' levels, which emphasizes the need of an individual profiling over time, to be able to detect meaningful variations and undesirable emotional or distress responses in each AD dog due to its particular training/daily life. The lack of correlation between the blood levels of these neuromodulators requires independent investigation of each of them to be able to draw an overall emotional profile. Ideally, an individual follow-up during all their careers of AD will allow to highlight if their neurohormonal profiles can be associated with their future success and if this profiling can be helpful in predicting it. Eventually, neurohormonal profiling may help improve the selection of smarter ADs as well as the assessment of their educational program and workload to guarantee the best possible success of the AD training.

The low correlation between fOT and tOT, as well as the biological meaningfulness of tOT warrants further investigation. Other neuromodulators may also show differences between AD and PD, and it would be interesting to explore these deeper in the future. In particular, vasopressin, the neuropeptide closely related to OT and also involved in a wide range of social behaviors, including aggression, and stress responses is certainly of high interest in this context [61]. Finally, it may also be beneficial to apply such approaches to other working dogs. Further studies improving our knowledge about their physiology, neuroendocrinology, and behavior, due to meaningful biomarkers to support desirable temperament/behavior and breed selection, will be surely welcomed to improve the training procedures and dogs' welfare.

5. Conclusions

In conclusion, this study established neurohormonal profiles of dogs (AD and PD) in order to give cues about their emotional and stress responses and coping abilities. This can be especially beneficial in the context of working dogs, such as assistance dogs in this study, to (i) improve their selection based on their inherent neurobiological features and avoid time and money loss by educating and training unsuitable dogs; and (ii) assess the suitability of their training program and their resulting lifestyle by evaluating their impact on dogs' stress, emotions, and welfare via the measurement of neurohormones linked to these processes. This study showed that AD can display different levels of various neurohormones than PD, particularly fOT and tOT, which are of high interest for AD, considering its involvement in bonding processes and social behaviors. To a lesser extent, PRL levels may also differ, and this should be further investigated, considering its association with emotional disorders. Finally, this study also emphasized the age influence on basal neurohormonal levels among dog populations. Thus, individual monitoring of neuromodulators over time, allowing to control for sex and age variability, seems to be preferable to properly evaluate each dog's emotional responses.

We hope this work can pave the way to a new approach based on individual neurohormones profiling/monitoring to assess the welfare states of dogs involved in animal-assisted activities and, in the end, improve their selection/training for better chances of success.

Author Contributions: Conceptualization, M.M. and C.B.-F.; data curation, C.C., S.A. and C.L.-L.; formal analysis, S.A. and C.L.-L.; funding acquisition, P.P.; investigation, M.M., J.L.O., T.M. and C.C.; methodology, M.M., J.L.O., C.L.-L. and C.B.-F.; project administration, A.C. and C.B.-F.; resources, M.M., J.L.O., T.M., C.C., A.C., P.P. and C.B.-F.; supervision, M.M. and C.B.-F.; visualization, M.M., S.A. and C.B.-F.; writing—original draft, M.M. and C.B.-F.; writing—review & editing, J.L.O., T.M., S.A., A.C., P.P. and C.B.-F. All authors have read and agreed to the published version of the manuscript.

Funding: This research received no external funding.

Institutional Review Board Statement: This study was performed in accordance with French law (2013-118) and European directive (2010/63/EU) on the protection of animals used for scientific purposes. The study was approved by the IRSEA Ethics Committee C2EA125 (approval number AFCE_201605_02).

Informed Consent Statement: Informed consent was obtained from all dogs' owners involved in the study.

Data Availability Statement: The data presented in this study are available on request from the corresponding author.

Acknowledgments: We thank Muriel Jochem and the Frédéric Gaillanne Foundation with all its members for their support and help all along the trial. We are very grateful to Eva Teruel and Estelle Descout from the IRSEA Data Management and Statistics Service for their help in the statistical analyses following the reviewing processes and we also sincerely thank the reviewers for their useful suggestions. We would also like to thank all the foster families and all the pet dogs' owners involved in this study for their help, their availability, and their interest in this trial.

Conflicts of Interest: The authors declare no conflict of interest.

References

1. Bremhorst, A.; Mongillo, P.; Howell, T.; Marinelli, L. Spotlight on assistance dogs—legislation, welfare and research. *Animals* **2018**, *8*, 129. [CrossRef]
2. Brady, K.; Cracknell, N.; Zulch, H.; Mills, D.S. Factors associated with long-term success in working police dogs. *Appl. Anim. Behav. Sci.* **2018**, *207*, 67–72. [CrossRef]
3. Gazzano, A.; Mariti, C.; Sighieri, C.; Ducci, M.; Ciceroni, C.; McBride, E.A. Survey of undesirable behaviors displayed by potential guide dogs with puppy walkers. *J. Vet. Behav. Clin. Appl. Res.* **2008**, *3*, 104–113. [CrossRef]
4. Koda, N.; Kubo, M.; Ishigami, T.; Furuhashi, H. Assessment of Dog Guides by Users in Japan and Suggestions for Improvement. *J. Vis. Impair. Blind.* **2011**, *105*, 591–600. [CrossRef]
5. Naderi, S.; Miklósi, Á.; Dóka, A.; Csányi, V. Co-operative interactions between blind persons and their dogs. *Appl. Anim. Behav. Sci.* **2001**, *74*, 59–80. [CrossRef]
6. Arata, S.; Momozawa, Y.; Takeuchi, Y.; Mori, Y. Important Behavioral Traits for Predicting Guide Dog Qualification. *J. Vet. Med. Sci.* **2010**, *72*, 539–545. [CrossRef] [PubMed]
7. Batt, L.S.; Batt, M.S.; Baguley, J.A.; McGreevy, P.D. Factors associated with success in guide dog training. *J. Vet. Behav. Clin. Appl. Res.* **2008**, *3*, 143–151. [CrossRef]
8. Menuge, F.; Marcet-Rius, M.; Jochem, M.; François, O.; Assali, C.; Chabaud, C.; Teruel, E.; Guillemot, J.; Pageat, P. Early evaluation of fearfulness in future guide dogs for blind people. *Animals* **2021**, *11*, 412. [CrossRef]
9. Serpell, J.A.; Hsu, Y. Development and validation of a novel method for evaluating behavior and temperament in guide dogs. *Appl. Anim. Behav. Sci.* **2001**, *72*, 347–364. [CrossRef]
10. Cobb, M.; Branson, N.; McGreevy, P.; Lill, A.; Bennett, P. The advent of canine performance science: Offering a sustainable future for working dogs. *Behav. Process.* **2015**, *110*, 96–104. [CrossRef]
11. Mengoli, M.; Mendonça, T.; Oliva, J.L.; Bienboire-Frosini, C.; Chabaud, C.; Codecasa, E. Do assistance dogs show work overload. In *Canine Blood Prolactin as a Clinical Parameter to Detect Chronic Stress-Related Response, Proceedings of the Proceedings of the 11th International Veterinary Behaviour Meeting, Samorin, Slovakia, 14–16 September 2017*; CABI: Wallingford, UK, 2017; Volume 45, p. 112.
12. Freeman, M.E.; Kanyicska, B.; Lerant, A.; Nagy, G. Prolactin: Structure, function, and regulation of secretion. *Physiol. Rev.* **2000**, *80*, 1703–1726. [CrossRef] [PubMed]
13. Landgraf, R.; Wigger, A.; Holsboer, F.; Neumann, I.D. Hyper-reactive hypothalamo-pituitary-adrenocortical axis in rats bred for high anxiety-related behaviour. *J. Neuroendocrinol.* **1999**, *11*, 405–407. [CrossRef] [PubMed]
14. Yayou, K.-I.; Ito, S.; Yamamoto, N.; Kitagawa, S.; Okamura, H. Relationships of stress responses with plasma oxytocin and prolactin in heifer calves. *Physiol. Behav.* **2010**, *99*, 362–369. [CrossRef] [PubMed]
15. Kataria, N.; Kataria, A.K. Use of prolactin as an indicator of stress in Marwari sheep from arid tracts in India. *Comp. Clin. Path.* **2011**, *20*, 333–336. [CrossRef]
16. Jeffcoate, W.J.; Lincoln, N.B.; Selby, C.; Herbert, M. Correlation between anxiety and serum prolactin in humans. *J. Psychosom. Res.* **1986**, *30*, 217–222. [CrossRef]
17. Turner, R.A.; Altemus, M.; Yip, D.N.; Kupferman, E.; Fletcher, D.; Bostrom, A.; Lyons, D.M.; Amico, J.A. Effects of emotion on oxytocin, prolactin, and ACTH in women. *Stress* **2002**, *5*, 269–276. [CrossRef] [PubMed]
18. Lennartsson, A.K.; Jonsdottir, I.H. Prolactin in response to acute psychosocial stress in healthy men and women. *Psychoneuroendocrinology* **2011**, *36*, 1530–1539. [CrossRef]
19. Pageat, P.; Lafont, C.; Falewée, C.; Bonnafous, L.; Gaultier, E.; Silliart, B. An evaluation of serum prolactin in anxious dogs and response to treatment with selegiline or fluoxetine. *Appl. Anim. Behav. Sci.* **2007**, *105*, 342–350. [CrossRef]
20. Gutiérrez, J.; Gazzano, A.; Pirrone, F.; Sighieri, C.; Mariti, C. Investigating the role of prolactin as a potential biomarker of stress in castrated male domestic dogs. *Animals* **2019**, *9*, 676. [CrossRef]
21. Oliva, J.L.; Mengoli, M.; Mendonça, T.; Cozzi, A.; Pageat, P.; Chabaud, C.; Teruel, E.; Lafont-Lecuelle, C.; Bienboire-Frosini, C. Working Smarter Not Harder: Oxytocin Increases Domestic Dogs' (*Canis familiaris*) Accuracy, but Not Attempts, on an Object Choice Task. *Front. Psychol.* **2019**, *10*, 1–18. [CrossRef] [PubMed]
22. Mosienko, V.; Beis, D.; Pasqualetti, M.; Waider, J.; Matthes, S.; Qadri, F.; Bader, M.; Alenina, N. Life without brain serotonin: Reevaluation of serotonin function with mice deficient in brain serotonin synthesis. *Behav. Brain Res.* **2015**, *277*, 78–88. [CrossRef]
23. Scott, K.A.; Tamashiro, K.L.K.; Sakai, R.R. Chronic social stress: Effects on neuroendocrine function. In *Handbook of Neuroendocrinology*; Fink, G., Pfaff, D.W., Levine, J.E., Eds.; Elsevier: Amsterdam, The Netherlands, 2012; pp. 521–534.
24. León, M.; Rosado, B.; García-Belenguer, S.; Chacón, G.; Villegas, A.; Palacio, J. Assessment of serotonin in serum, plasma, and platelets of aggressive dogs. *J. Vet. Behav. Clin. Appl. Res.* **2012**, *7*, 348–352. [CrossRef]
25. Amat, M.; Le Brech, S.; Camps, T.; Torrente, C.; Mariotti, V.M.; Ruiz, J.L.; Manteca, X. Differences in serotonin serum concentration between aggressive English cocker spaniels and aggressive dogs of other breeds. *J. Vet. Behav. Clin. Appl. Res.* **2013**, *8*, 19–25. [CrossRef]
26. Çakiroğlu, D.; Meral, Y.; Sancak, A.A.; Çifti, G. Relationship between the serum concentrations of serotonin and lipids and aggression in dogs. *Vet. Rec.* **2007**, *161*, 59–61. [CrossRef]
27. Gimpl, G.; Fahrenholz, F. The Oxytocin Receptor System: Structure, Function, and Regulation. *Physiol. Rev.* **2001**, *81*, 629–683. [CrossRef] [PubMed]

28. Lim, M.M.; Young, L.J. Neuropeptidergic regulation of affiliative behavior and social bonding in animals. *Horm. Behav.* **2006**, *50*, 506–517. [CrossRef] [PubMed]
29. Lee, H.-J.; Macbeth, A.H.; Pagani, J.H.; Young, W.S. Oxytocin: The great facilitator of life. *Prog. Neurobiol.* **2009**, *88*, 127–151. [CrossRef] [PubMed]
30. Ziegler, T.E.; Crockford, C. Neuroendocrine control in social relationships in non-human primates: Field based evidence. *Horm. Behav.* **2017**, *91*, 107–121. [CrossRef]
31. Bosch, O.J.; Neumann, I.D. Both oxytocin and vasopressin are mediators of maternal care and aggression in rodents: From central release to sites of action. *Horm. Behav.* **2012**, *61*, 293–303. [CrossRef]
32. Ditzen, B.; Schaer, M.; Gabriel, B.; Bodenmann, G.; Ehlert, U.; Heinrichs, M. Intranasal Oxytocin Increases Positive Communication and Reduces Cortisol Levels During Couple Conflict. *Biol. Psychiatry* **2009**, *65*, 728–731. [CrossRef]
33. Dunbar, R.I.M. The social role of touch in humans and primates: Behavioural function and neurobiological mechanisms. *Neurosci. Biobehav. Rev.* **2010**, *34*, 260–268. [CrossRef]
34. Romero, T.; Nagasawa, M.; Mogi, K.; Hasegawa, T.; Kikusui, T. Intranasal administration of oxytocin promotes social play in domestic dogs. *Commun. Integr. Biol.* **2015**, *8*, 20–23. [CrossRef] [PubMed]
35. Numan, M.; Young, L.J. Neural mechanisms of mother-infant bonding and pair bonding: Similarities, differences, and broader implications. *Horm. Behav.* **2016**, *77*, 98–112. [CrossRef]
36. Odendaal, J.S.J.; Meintjes, R.A. Neurophysiological Correlates of Affiliative Behaviour between Humans and Dogs. *Vet. J.* **2003**, *165*, 296–301. [CrossRef]
37. Beetz, A.; Uvnäs-Moberg, K.; Julius, H.; Kotrschal, K. Psychosocial and psychophysiological effects of human-animal interactions: The possible role of oxytocin. *Front. Psychol.* **2012**, *3*, 234. [CrossRef]
38. Romero, T.; Nagasawa, M.; Mogi, K.; Hasegawa, T.; Kikusui, T. Oxytocin promotes social bonding in dogs. *Proc. Natl. Acad. Sci. USA* **2014**, *111*, 9085–9090. [CrossRef]
39. Rault, J.L. Effects of positive and negative human contacts and intranasal oxytocin on cerebrospinal fluid oxytocin. *Psychoneuroendocrinology* **2016**, *69*, 60–66. [CrossRef] [PubMed]
40. Lürzel, S.; Bückendorf, L.; Waiblinger, S.; Rault, J.L. Salivary oxytocin in pigs, cattle, and goats during positive human-animal interactions. *Psychoneuroendocrinology* **2020**, *115*. [CrossRef] [PubMed]
41. Ogi, A.; Mariti, C.; Baragli, P.; Sergi, V.; Gazzano, A. Effects of stroking on salivary oxytocin and cortisol in guide dogs: Preliminary results. *Animals* **2020**, *10*, 708. [CrossRef] [PubMed]
42. MacLean, E.L.; Gesquiere, L.R.; Gee, N.; Levy, K.; Martin, W.L.; Carter, C.S. Effects of Affiliative Human-Animal Interaction on Dog Salivary and Plasma Oxytocin and Vasopressin. *Front. Psychol.* **2017**, *8*, 1–9. [CrossRef]
43. Ogi, A.; Torracca, B.; Gutierrez, J.; Mariti, C.; Gazzano, A. Effects of human-animal interaction and isolation on salivary oxytocin in 8 guide dogs. In *Proceedings of the 1st European Congress of Veterinary Behavioural Medicine and Animal Welfare, Berlin, Germany, 27–29 September 2018*; Verlag: Giessen, Germany, 2018; pp. 290–292.
44. Kis, A.; Hernadi, A.; Kanizsar, O.; Gacsi, M.; Topál, J. Oxytocin induces positive expectations about ambivalent stimuli (cognitive bias) in dogs. *Horm. Behav.* **2015**, *69*, 1–7. [CrossRef]
45. Kis, A.; Hernádi, A.; Miklósi, B.; Kanizsár, O.; Topál, J. The way dogs (*Canis familiaris*) look at human emotional faces is modulated by oxytocin. an eye-tracking study. *Front. Behav. Neurosci.* **2017**, *11*. [CrossRef] [PubMed]
46. Dzik, M.V.; Cavalli, C.M.; Barrera, G.; Bentosela, M. Oxytocin effects on gazing at the human face in retriever dogs. *Behav. Process.* **2020**, *178*, 104160. [CrossRef]
47. Dzik, M.V.; Carballo, F.; Casanave, E.; Bentosela, M. Effects of oxytocin administration and the dog–owner bond on dogs' rescue behavior. *Anim. Cogn.* **2021**, 1–14. [CrossRef]
48. Oliva, J.L.; Rault, J.L.; Appleton, B.; Lill, A. Oxytocin enhances the appropriate use of human social cues by the domestic dog (Canis familiaris) in an object choice task. *Anim. Cogn.* **2015**, *18*, 767–775. [CrossRef] [PubMed]
49. Hernadi, A.; Kis, A.; Kanizsár, O.; Tóth, K.; Miklósi, B.; Topál, J. Intranasally administered oxytocin affects how dogs (*Canis familiaris*) react to the threatening approach of their owner and an unfamiliar experimenter. *Behav. Process.* **2015**, *119*, 1–5. [CrossRef] [PubMed]
50. Bienboire-Frosini, C.; Chabaud, C.; Mendonça, T.; Mengoli, M.; Arroub, S.; Oliva, J.L.; Cozzi, A.; Pageat, P. Investigating the oxytocinergic system: Plasma measures of free vs. total oxytocin fractions in two animal models. *Psychoneuroendocrinology* **2019**, *107*, 33. [CrossRef]
51. Marshall-Pescini, S.; Schaebs, F.S.; Gaugg, A.; Meinert, A.; Deschner, T.; Range, F. The role of oxytocin in the dog–owner relationship. *Animals* **2019**, *9*, 792. [CrossRef]
52. Wotjak, C.T.; Ganster, J.; Kohl, G.; Holsboer, F.; Landgraf, R.; Engelmann, M. Dissociated central and peripheral release of vasopressin, but not oxytocin, in response to repeated swim stress: New insights into the secretory capacities of peptidergic neurons. *Neuroscience* **1998**, *85*, 1209–1222. [CrossRef]
53. Ross, H.E.; Young, L.J. Oxytocin and the neural mechanisms regulating social cognition and affiliative behavior. *Front. Neuroendocrinol.* **2009**, *30*, 534–547. [CrossRef]
54. Carson, D.S.; Garner, J.P.; Hyde, S.A.; Libove, R.A.; Berquist, S.W.; Hornbeak, K.B.; Jackson, L.P.; Sumiyoshi, R.D.; Howerton, C.L.; Hannah, S.L.; et al. Arginine vasopressin is a blood-based biomarker of social functioning in children with autism. *PLoS ONE* **2015**, *10*. [CrossRef]

55. Bienboire-Frosini, C.; Chabaud, C.; Cozzi, A.; Codecasa, E.; Pageat, P. Validation of a commercially available enzyme immunoassay for the determination of oxytocin in plasma samples from seven domestic animal species. *Front. Neurosci.* **2017**, *11*. [CrossRef]
56. MacLean, E.L.; Gesquiere, L.R.; Gee, N.; Levy, K.; Martin, W.L.; Carter, C.S. Validation of salivary oxytocin and vasopressin as biomarkers in domestic dogs. *J. Neurosci. Methods* **2018**, *293*, 67–76. [CrossRef]
57. Schaebs, F.S.; Marshall-Pescini, S.; Range, F.; Deschner, T. Analytical validation of an Enzyme Immunoassay for the measurement of urinary oxytocin in dogs and wolves. *Gen. Comp. Endocrinol.* **2019**. [CrossRef] [PubMed]
58. Carter, C.S.; Kenkel, W.M.; Maclean, E.L.; Wilson, S.R.; Perkeybile, A.M.; Yee, J.R.; Ferris, C.F.; Nazarloo, H.P.; Porges, S.W.; Davis, J.M.; et al. Is oxytocin "nature's medicine"? *Pharmacol. Rev.* **2020**, *72*, 829–861. [CrossRef] [PubMed]
59. MacLean, E.L.; Wilson, S.R.; Martin, W.L.; Davis, J.M.; Nazarloo, H.P.; Carter, C.S. Challenges for measuring oxytocin: The blind men and the elephant? *Psychoneuroendocrinology* **2019**, *107*, 225–231. [CrossRef] [PubMed]
60. Brandtzaeg, O.K.; Johnsen, E.; Roberg-larsen, H.; Seip, K.F.; Maclean, E.L.; Gesquiere, L.R.; Leknes, S.; Lundanes, E.; Wilson, S.R. Proteomics tools reveal startlingly high amounts of oxytocin in plasma and serum. *Sci. Rep.* **2016**, 1–7. [CrossRef] [PubMed]
61. MacLean, E.L.; Gesquiere, L.R.; Gruen, M.E.; Sherman, B.L.; Martin, W.L.; Carter, C.S. Endogenous oxytocin, vasopressin, and aggression in domestic dogs. *Front. Psychol.* **2017**, *8*. [CrossRef]
62. López-Arjona, M.; Mateo, S.V.; Escribano, D.; Tecles, F.; Cerón, J.J.; Martínez-Subiela, S. Effect of reduction and alkylation treatment in three different assays used for the measurement of oxytocin in saliva of pigs. *Domest. Anim. Endocrinol.* **2021**, *74*. [CrossRef] [PubMed]
63. Leng, G.; Ludwig, M. Intranasal Oxytocin: Myths and Delusions. *Biol. Psychiatry* **2016**, *79*, 243–250. [CrossRef]
64. Crockford, C.; Deschner, T.; Ziegler, T.E.; Wittig, R.M. Endogenous peripheral oxytocin measures can give insight into the dynamics of social relationships: A review. *Front. Behav. Neurosci.* **2014**, *8*, 68. [CrossRef] [PubMed]
65. Moberg, K.U.; Handlin, L.; Kendall-Tackett, K.; Petersson, M. Oxytocin is a principal hormone that exerts part of its effects by active fragments. *Med. Hypotheses* **2019**, *133*. [CrossRef]
66. Kennett, J.E.; McKee, D.T. Oxytocin: An emerging regulator of prolactin secretion in the female rat. *J. Neuroendocrinol.* **2012**, *24*, 403–412. [CrossRef]
67. Onaka, T.; Takayanagi, Y.; Yoshida, M. Roles of oxytocin neurones in the control of stress, energy metabolism, and social behaviour. *J. Neuroendocrinol.* **2012**, *24*, 587–598. [CrossRef] [PubMed]
68. Chabaud, C.; Mathieu, M.; Brooks, E.; Bienboire-Frosini, C. Application note: Validation of a Serotonin ELISA Kit with Blood Samples from Three Domestic Animal Species. Available online: https://www.enzolifesciences.com/about-us/collaborations-at-work/immunoassay/validation-of-a-serotonin-elisa-kit-with-blood-samples-from-three-domestic-animal-species/ (accessed on 30 April 2021).
69. Park, H.J.; Lee, S.E.; Oh, J.H.; Seo, K.W.; Song, K.H. Leptin, adiponectin and serotonin levels in lean and obese dogs. *BMC Vet. Res.* **2014**, *10*, 1–8. [CrossRef]
70. Gobello, C.; Brown, O.; Rondero, J. Determinaciones hormonales. *Analecta Vet.* **1998**, *18*, 71–81.
71. Siracusa, C.; Manteca, X.; Cuenca, R.; del Mar Alcalá, M.; Alba, A.; Lavín, S.; Pastor, J. Effect of a synthetic appeasing pheromone on behavioral, neuroendocrine, immune, and acute-phase perioperative stress responses in dogs. *J. Am. Vet. Med. Assoc.* **2010**, *237*, 673–681. [CrossRef] [PubMed]
72. Corrada, Y.; Rimoldi, I.; Arreseigor, S.; Marecco, G.; Gobello, C. Prolactin reference range and pulsatility in male dogs. *Theriogenology* **2006**, *66*, 1599–1602. [CrossRef]
73. Urhausen, C.; Seefeldt, A.; Eschricht, F.; Koch, A.; Hoppen, H.; Beyerbach, M.; Möhrke, C.; Dieleman, S.; Günzel-Apel, A. Concentrations of Prolactin, LH, Testosterone, TSH and Thyroxine in Normospermic Dogs of Different Breeds. *Reprod. Domest. Anim.* **2009**, *44*, 279–282. [CrossRef]
74. Ben-Jonathan, N.; Hnasko, R. Dopamine as a prolactin (PRL) inhibitor. *Endocr. Rev.* **2001**, *22*, 724–763. [CrossRef]
75. Osella, M.C.; Odore, R.; Badino, P.; Cuniberti, B.; Bergamasco, L. Plasma dopamine neurophysiological correlates in anxious dogs. In *Current Issues and Research in Veterinary Behavioral Medicine*; Purdue University Press: West Lafayette, IN, USA, 2005; pp. 274–276.
76. MacGregor, D.A.; Smith, T.E.; Prielipp, R.C.; Butterworth, J.F.; James, R.L.; Scuderi, P.E. Pharmacokinetics of Dopamine in Healthy Male Subjects. *Anesthesiology* **2000**, *92*, 338. [CrossRef]
77. McDannald, M.A. Serotonin: Waiting but not rewarding. *Curr. Biol.* **2015**, *25*, R103–R104. [CrossRef]
78. Carhart-Harris, R.L.; Nutt, D.J. Serotonin and brain function: A tale of two receptors. *J. Psychopharmacol.* **2017**, *31*, 1091–1120. [CrossRef]
79. Alberghina, D.; Rizzo, M.; Piccione, G.; Giannetto, C.; Panzera, M. An exploratory study about the association between serum serotonin concentrations and canine-human social interactions in shelter dogs (*Canis familiaris*). *J. Vet. Behav. Clin. Appl. Res.* **2017**, *18*, 96–101. [CrossRef]
80. Alberghina, D.; Tropia, E.; Piccione, G.; Giannetto, C.; Panzera, M. Serum serotonin (5-HT) in dogs (*Canis familiaris*): Preanalytical factors and analytical procedure for use of reference values in behavioral medicine. *J. Vet. Behav.* **2019**, *32*, 72–75. [CrossRef]
81. Höglund, K.; Häggström, J.; Hanås, S.; Merveille, A.C.; Gouni, V.; Wiberg, M.; Lundgren Willesen, J.; Entee, K.M.; Mejer Sørensen, L.; Tiret, L.; et al. Interbreed variation in serum serotonin (5-hydroxytryptamine) concentration in healthy dogs. *J. Vet. Cardiol.* **2018**, *20*, 244–253. [CrossRef] [PubMed]

82. Arndt, J.; Reynolds, C.; Singletary, G.; Connolly, J.; Levy, R.; Oyama, M. Serum Serotonin Concentrations in Dogs with Degenerative Mitral Valve Disease. *J. Vet. Intern. Med.* **2009**, *23*, 1208–1213. [CrossRef] [PubMed]
83. Alberghina, D.; Piccione, G.; Lombardo, G.; Taormina, A.; Panzera, M. Serun serotonin levels in dogs: Influence of two different housing environments. *J. Vet. Behav.* **2014**, *9*, e1. [CrossRef]
84. Karakulova, U.V.; Selyanina, N.V.; Sumlivaya, O.N.; Vorobyeva, N.N.; Okishev, M.A. Involvement of peripheral unit of serotonin energetic system in brain lesions of different types. *World Appl. Sci. J.* **2013**, *26*, 374–376. [CrossRef]
85. Kovács, K.; Kis, A.; Pogány, Á.; Koller, D.; Topál, J. Differential effects of oxytocin on social sensitivity in two distinct breeds of dogs (*Canis familiaris*). *Psychoneuroendocrinology* **2016**, *74*, 212–220. [CrossRef] [PubMed]
86. Nagasawa, M.; Mitsui, S.; En, S.; Ohtani, N.; Ohta, M.; Sakuma, Y.; Onaka, T.; Mogi, K.; Kikusui, T. Oxytocin-gaze positive loop and the coevolution of human-dog bonds. *Science* **2015**, *348*, 333–336. [CrossRef] [PubMed]
87. Oliva, J.L.; Wong, Y.T.; Rault, J.-L.; Appleton, B.; Lill, A. The oxytocin receptor gene, an integral piece of the evolution of *Canis familaris* from Canis lupus. *Pet. Behav. Sci.* **2016**, *2*, 1. [CrossRef]
88. Wirobski, G.; Range, F.; Schaebs, F.S.; Palme, R.; Deschner, T.; Pescini, S.M. Life experience rather than domestication accounts for dogs ' increased oxytocin release during social contact with humans. *Sci. Rep.* **2021**, 1–12. [CrossRef]
89. Yoshida, M.; Takayanagi, Y.; Inoue, K.; Kimura, T.; Young, L.J.; Onaka, T.; Nishimori, K. Evidence that oxytocin exerts anxiolytic effects via oxytocin receptor expressed in serotonergic neurons in mice. *J. Neurosci.* **2009**, *29*, 2259–2271. [CrossRef]
90. Dölen, G.; Darvishzadeh, A.; Huang, K.W.; Malenka, R.C. Social reward requires coordinated activity of nucleus accumbens oxytocin and serotonin. *Nature* **2013**, *501*, 179–184. [CrossRef]
91. Mottolese, R.; Redouté, J.; Costes, N.; Le Bars, D.; Sirigu, A. Switching brain serotonin with oxytocin. *Proc. Natl. Acad. Sci. USA* **2014**, *111*, 8637–8642. [CrossRef] [PubMed]
92. Patterson-Kane, E.; Yamamoto, M.; Hart, L.A. Editorial: Assistance Dogs for People With Disabilities. *Front. Vet. Sci.* **2020**, *7*, 1–2. [CrossRef] [PubMed]

Review

Current Advances in Assessment of Dog's Emotions, Facial Expressions, and Their Use for Clinical Recognition of Pain

Daniel Mota-Rojas [1,*], Míriam Marcet-Rius [2], Asahi Ogi [3], Ismael Hernández-Ávalos [4], Chiara Mariti [3], Julio Martínez-Burnes [5], Patricia Mora-Medina [6], Alejandro Casas [1], Adriana Domínguez [1], Brenda Reyes [1] and Angelo Gazzano [3]

1 Neurophysiology of Pain, Behavior and Animal Welfare Assessment, DPAA, Universidad Autónoma Metropolitana (UAM), Mexico City 04960, Mexico; ale0164g@hotmail.com (A.C.); mvz.freena@gmail.com (A.D.); breyess_20@yahoo.com.mx (B.R.)
2 Animal Behaviour and Welfare Department, IRSEA (Research Institute in Semiochemistry and Applied Ethology), Quartier Salignan, 84400 Apt, France; m.marcet@group-irsea.com
3 Department of Veterinary Sciences, University of Pisa, 56124 Pisa, Italy; asahi.ogi@vet.unipi.it (A.O.); chiara.mariti@unipi.it (C.M.); angelo.gazzano@unipi.it (A.G.)
4 Department of Biological Sciences, Clinical Pharmacology and Veterinary Anaesthesia, FESC, Universidad Nacional Autónoma de México (UNAM), Cuautitlán Izcalli 54714, Mexico; mvziha@hotmail.com
5 Animal Health Group, Facultad de Medicina Veterinaria y Zootecnia, Universidad Autónoma de Tamaulipas, Victoria City 87000, Mexico; jmburnes@docentes.uat.edu.mx
6 Department of Livestock Science, FESC, Universidad Nacional Autónoma de México (UNAM), Cuautitlán Izcalli 54714, Mexico; mormed2001@yahoo.com.mx
* Correspondence: dmota100@yahoo.com.mx

Simple Summary: In several species, facial expressions have been associated with positive and negative emotions to communicate their mental state. In dogs, the interpretation of these muscle movements is relevant because of their close bond with humans. Currently, there is a discussion about whether facial expressions in domestic dogs can communicate emotions or are simply the result of mimicry and emotional contagion. This article will discuss the available literature on dogs' facial expressions, anatomy and neurophysiology, and their association with emotions and adverse events such as pain. In this species, it is a challenge to identify and associate both factors due to domestication. This review aims to provide scientific support and understanding of facial expression in dogs as a clinical ethological tool.

Abstract: Animals' facial expressions are involuntary responses that serve to communicate the emotions that individuals feel. Due to their close co-existence with humans, broad attention has been given to identifying these expressions in certain species, especially dogs. This review aims to analyze and discuss the advances in identifying the facial expressions of domestic dogs and their clinical utility in recognizing pain as a method to improve daily practice and, in an accessible and effective way, assess the health outcome of dogs. This study focuses on aspects related to the anatomy and physiology of facial expressions in dogs, their emotions, and evaluations of their eyebrows, eyes, lips, and ear positions as changes that reflect pain or nociception. In this regard, research has found that dogs have anatomical configurations that allow them to generate changes in their expressions that similar canids—wolves, for example—cannot produce. Additionally, dogs can perceive emotions similar to those of their human tutors due to close human-animal interaction. This phenomenon—called "emotional contagion"—is triggered precisely by the dog's capacity to identify their owners' gestures and then react by emitting responses with either similar or opposed expressions that correspond to positive or negative stimuli, respectively. In conclusion, facial expressions are essential to maintaining social interaction between dogs and other species, as in their bond with humans. Moreover, this provides valuable information on emotions and the perception of pain, so in dogs, they can serve as valuable elements for recognizing and evaluating pain in clinical settings.

Keywords: animal welfare; emotions; human-dog interaction; pain; positive and negative stimuli

Citation: Mota-Rojas, D.; Marcet-Rius, M.; Ogi, A.; Hernández-Ávalos, I.; Mariti, C.; Martínez-Burnes, J.; Mora-Medina, P.; Casas, A.; Domínguez, A.; Reyes, B.; et al. Current Advances in Assessment of Dog's Emotions, Facial Expressions, and Their Use for Clinical Recognition of Pain. *Animals* **2021**, *11*, 3334. https://doi.org/10.3390/ani11113334

Academic Editor: Lee Niel

Received: 11 September 2021
Accepted: 19 November 2021
Published: 22 November 2021

1. Introduction

Identifying facial expressions in animals has been relevant [1,2] since Darwin [3] stated that non-human animals can create innate expressions adaptable to each species. However, its purpose is still debatable. Whether they function as a non-verbal language to maintain social structure or to convey emotional states (neurophysiological changes associated with the recognition of pleasant and unpleasant emotions), in this sense, both emotional states and facial expressions require the integration of peripheral, autonomic, endocrine, and muscular responses, which involve the activation of various brain structures (i.e., the amygdala, hypothalamus, and brainstem) [4–7]. The understanding of the interaction between these responses and its neural pathways has led to studying facial expressions as a way to assess animal welfare through the identification of pain or stressful states in horses [8,9], sheep [10,11], laboratory animals [2], cows [12] and pigs [13–16].

In contrast, evolutionary and anatomical adaptations have been reported in domestic canines [17]. An example of these changes is raising the eyebrow, conferred by facial muscles only present in the dog, compared to wolves. These characteristics bestow a childish look or paedomorphic traits to this species [18].

The close relationship that dogs have with humans and the effects of domestication [19] has led domestic canines to develop the ability to detect, distinguish and respond to conspecific gestures [20–22]. Likewise, it has contributed to developing communication skills and interpreting their emotions, where emotions describe an internal state modulated by the central nervous system, in which physiological, behavioral, and cognitive mechanisms develop in response to a stimulus or event. Additionally, this phenomenon is considered as a form of empathy towards other individuals in their social circle [23]. The above supports the theory that facial expression plays an emotional role where identifying these visual signals is a means of emotional communication [24]. The idea can also be reinforced with the so-called "emotional contagion", a process in which animals learn to identify expressions from the same or other species and react with a similar pattern [22].

In this way, research regarding this topic has associated lip or eyebrow movements with positive and negative emotional states (known as the state caused by emotions such as fear, pain, or disgust) during heterospecific relationships with humans, and these can provide information about the emotions perceived by dogs [25]. Given this scenario, the evaluation of facial expressions has a relevant clinical value for pain diagnosis [26] and can be used together with current pain assessments and recognition [27]. As mentioned by Häger et al. [11], conscious perception of pain is represented by a change in facial expressions, such as ears flattening and tension in the muscles of the nose, mouth, and the orbital region [28,29]. Identifying these changes has been proposed as an alternative to assess the degree of pain through grimace scales based on the facial expression of cats [30,31], pigs [13], mice [32], and rats [33].

Thus, understanding the different processes that involve animals' emotions and how to assess them through facial expressions will surely be a valuable approach in medical fields of study, such as neuroscience, psychopharmacology, animal behavior, algology, and the science of animal welfare [34]. This review aims to discuss and analyze the current advances in identifying facial expressions of domestic dogs and their clinical utility to recognize emotions, including pain, as a method to improve daily practice and, in an accessible and effective way, assess dogs' health outcomes.

2. Methodology

A comprehensive literature review was conducted using *Web* of Science, Scopus, Science Direct, and PubMed. The keywords applied were related to "dog facial expression", "animal emotions", "dog facial action units", "evaluation of emotions in dogs", "validation of emotions", "facial neuroanatomy", "dog face anatomy", and "facial expression and pain". The inclusion criteria were articles published from 2000 to 2021, articles related to evaluating facial expressions in dogs and their emotions, the association between facial changes and pain perception, and pain conduction pathways and methods of identification and clinical recognition of pain. The exclusion criteria were articles related to other species and papers published before the year 2000 (except those that serve as a basis for understanding emotions or basic canine anatomy). Figure 1 describes the overall methodology for this review.

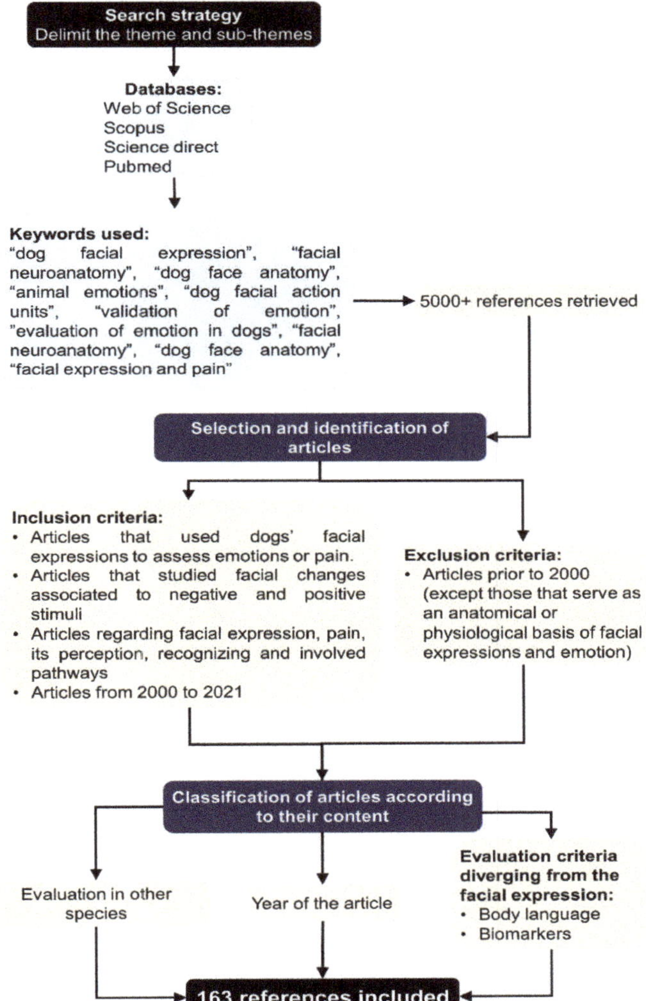

Figure 1. Review methodology.

3. Are Facial Expressions Involuntary and Emotional or Do They Have a Real Communication Purpose?

Darwin [3] argued that nonhuman animals communicate emotions through facial expressions. From the human perspective, canine facial expressions are considered a non-verbal communication tool not linked to the emotional concept [1]. This recent interpretation contrasts with the classical one, establishing that facial changes transmit information on six primary affective states: fear, happiness, anger, sadness, neutrality, and pain [35], which trigger specific responses, depending on the valence of the perceived emotion [36]. In animals, it has been suggested that facial expressions serve as a means of communication of emotional states [5]. Additionally, they also communicate sensations and intentions [37] to avoid predation and achieve a superior hierarchy [38].

In dogs, the domestication process has changed some facial features that confer a paedomorphic or child-like appearance (Figure 2). These characteristics lead to feasible analogies with humans to study the meaning of facial expressions in dogs [21].

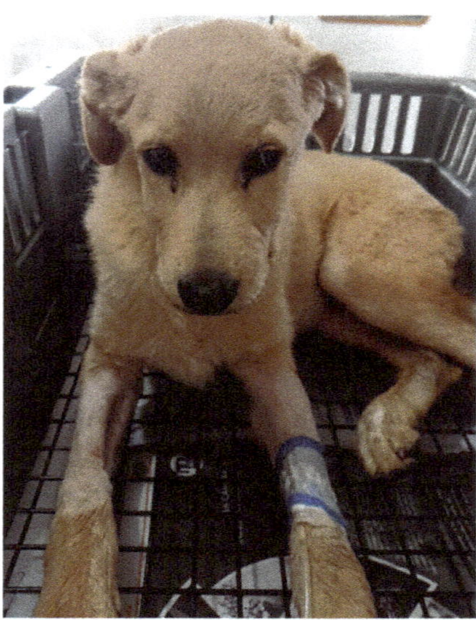

Figure 2. Characteristics of an infant-like, or paedomorphic face. A 3-month-old Golden Retriever puppy with gastrointestinal disease. The image shows the broad forehead and large eyes suggestive of sadness, two traits that may also be associated with chronic pain [39].

The study of facial expressions as a system of communication between conspecifics has been primarily investigated in dogs. Eyebrow and ear movements are used in dogs regularly to identify and communicate aversive states (referred to as an event caused by a noxious stimulus such as pain, distress, or any negative emotion), for example, flattening the ears or modifying the opening of the eyes during challenging stimuli [40]. In the same way, the movement of the lips, such as drawing forward the short lip and retraction of the labial commissure (in the long lip), communicate menacing states proportionally to the degree or intensity of the animal's reaction [41,42].

However, recent evidence has shown that facial expressions may not have an emotional role. Kaminski et al. [43] evaluated 24 dogs of different breeds to assess whether canine facial expressions are subject to the effects of the presence of humans or a positive stimulus such as food. The facial movements were greater in dogs when the person was attentive than when he or she was not paying attention. Moreover, facial movements were not affected when a non-social stimulus such as food was presented. The authors concluded that canine facial expressions are influenced by the presence of humans and their level of attention. These findings suggest a communicative function, rather than emotional, that can serve as a nonverbal language to express aversive responses such as pain [44].

In a study by Bremhorst et al. [45], 21 Labrador X Labrador retrievers were exposed to two scenarios to assess facial changes while waiting for a reward with the presence or absence of people (social and non-social context, respectively). In this work, there was a significant difference in the frequency of movement of the inner brow raiser, which was more frequent during the non-social stage ($p < 0.00001$, F = 24.62). This confronts the idea that facial expressions are produced by perceiving emotions, like ear and lip movements in negative conditions [46].

Another study by Bloom and Friedman [25] classified the facial expressions of dogs through photographs analyzed by humans (including experts in canine behavior, non-experts, and people who had lived with dogs for years) who identified facial movements related to happiness, sadness, surprise, disgust, anger, fear, and neutrality. Despite the results obtained, the study concluded that further research is required because the human interpretation of animal expression and emotion is highly subjective and can lead to mistakes. The lack of awareness of the relationship between facial movements, behavior and a specific emotional state can be a factor that contributes to this effect, leading to inaccurate interpretation of the animal's emotions. Therefore, it has been suggested that previous experience or knowledge about animal ethology influences the ability to recognize facial expressions. This subject will be addressed in detail later. Thus, the term "emotion" is a mental process that triggers reactions directed to specific internal and external stimuli which in animals is still controversial, since it is widely believed that emotions carry a subjective component that only humans can verbally express [47]. However, other authors differ and propose that emotions are physiological responses translated into corporal, vocal, and facial changes [48,49] that must be incorporated in future studies of animal emotion and behavior [5].

4. The Anatomy of Facial Expressions in Dogs

Currently, it is known that facial expressions are fundamental for the evolution and adaptation of species to their social environment and represent coordinated physiological and behavioral responses [36]. The standardized Facial Coding System (FACS) in animals is a detailed tool that identifies muscle movements using the so-called facial action units. These units identify and classify emotional states as neutral, fear, anger, and pleasure [25,50] based on the position and movement of the muscles. Movements such as the flattening of the ears, caused by the *inferior adductor auris, frontoscutilaris,* and the *retractor anguli occuli lateralis* muscles [46], as well as the nose wrinkles caused by lifting the upper lip due to the contraction of the *levator nasiolabialis,* as well as the *levator labii maxillaris* and *caninus,* are muscle movements associated with the perception of negative stimuli such as pain or fear in other species [51].

The facial nerve innervates and regulates the movement of the face through the cutaneous muscle of the neck (platysma), the zygomaticus, the buccinator, and the mentalis. This cranial nerve has both motor and sensory fibers afferents from the parasympathetic nervous system, regulated by the peripheral nervous system [52,53].

The facial nerve emerges from the cranial bone tissue of the stylomastoid foramen, located between the mastoid and styloid processes of the temporal bone. It innervates the occipital, posterior belly of the digastric, and stylohyoid muscles [54]. The facial nerve also leaves the skull and passes superficially to the parotid gland to originate its terminal branches in the cervical limits (temporal, zygomatic, buccal, marginal of the mandible, and cervical). These branches finally disperse and innervate the muscles of facial expression or facial mimicry [52,55].

The anatomy of the dog's face has a series of muscles that produce specific movements recognized in this species. For example, the platysma muscle is composed of longitudinally arranged fibers that allow the caudal retraction of the labial commissure, described as lip tightener (AU24) by the Dog Facial Coding System (DogFACS) [56]. This division includes the upper and lower muscles. Among the superiors, *the frontalis, retractor anguli occuli lateralis, levator anguli occuli medialis,* and *orbicularis occuli* allow movements such as a brow lift (AU101) and eyelid closure (AU 143) [20]. In the lower muscle group, the digastric muscle in its posterior belly, the *buccinator,* the *orbicularis oris, zygomaticus, mentalis, caninus, levator labii maxillaris,* and *levator nasiolabialis* originate movements in the ventral edge of the jaw to push it down (AU25), open the dog's mouth (AU17), deepen the nasolabial fold (AU11), or tighten the corners of the lips (AU27) [57–59]. Burrows et al. [60] mention that the marked anatomical differences in the faces of canids hinder reading their facial expressions because dogs have more than 20 facial movements; this is in comparison with other species, such as the chimpanzee or the cat, which only present 14 and 15 actions, respectively [61,62].

All canids have the same facial muscles, but beyond this anatomical similarity, it is essential to understand that the close co-existence of dogs with humans has developed some specific changes in dogs and breeds compared to wolves. Facial expressions in dogs allow them to manifest certain convenient gestures to maintain social structure [63]. One difference in dogs is the musculature responsible for eye movements, with significant development of the AU101 muscle. This muscle is also called the *levator anguli occuli medialis* (LAOM), or the interior elevator of the eyebrow. One of its functions is to give the impression of large eyes [18]. When it contracts, it enhances visual contact and confers a child-like appearance or paedomorphic features [18,64].

In the wolf, in contrast, the LAOM is small and surrounded by connective tissue, so the difference in movement in the AU101 region is greater in dogs [63]. This specific feature of the eyebrows in dogs is associated with dog-human interactions, which leads the latter to prefer certain breeds [65] (Figure 3). In comparison, observations of Siberian dogs, close relatives of the wolf, show little expressiveness in the eyes, because their muscular fibers are scarce and surrounded by large amounts of connective tissue, which impedes eyebrow movement. This internal muscular eyebrow movement in dogs is similar to human expressions of sadness, though it has also been associated with pain or malaise. It is worth mentioning that dogs developed specific movements due to these anatomical characteristics [63].

An example of this is the elevator of the eyebrow, which enlarges the eyes. Humans tend to associate this appearance with tenderness and care-motivation [65,66]. Although it has not been possible to demonstrate whether this characteristic represents a selection or preference advantage in humans [56], it is suggested that it may influence the interpretation of the mental state (a sensory state triggered by the perception of a specific stimulus). Understanding the emotional and facial repertoire is necessary to avoid misinterpretation of these movements, which are often mistaken as negative, such as sadness [67].

Another aspect to emphasize is that facial expressions involve more than one muscle. For example, the frontalis muscle, located in the frontal region between the scutiform cartilage and the upper eyelid, contracts in expressions of surprise and causes movement of the frontal region skin and the upper eyelid [68].

5. The Physiology of Facial Expressions

Animals' facial expressions have been linked to manifestations of emotions. Studies have determined that triggering a response to an external stimulus requires, first, decodifying that stimulus [69,70]. The response is then performed by the central nervous system (SNC), specifically by activation of the limbic system (the ventral and dorsal regions of the cingulate gyrus, prefrontal cortex, ventral striatum, dorsomedial nucleus of the thalamus and amygdala). This network is responsible for identifying motivations and processing emotions, while the amygdala generates emotions and the associated facial expressions [71]. It is also involved in emotional memory and the recognition of non-facial external stimuli, including stimuli of an auditory, olfactory or gustatory nature [72]. The communication between the amygdala and other structures of the limbic system is mediated by the secretion of neurotransmitters, neuropeptides, and hormones. These neurochemicals are involved in the presence of emotions such as fear, anxiety, and happiness, in which particularly high concentrations of catecholamines such as dopamine (DA), noradrenaline (NA), and adrenaline (A) can be found [73,74].

The release of oxytocin (OXT) facilitates emotional memory, which promotes social learning [75]. Negative emotions are attributed to these neurotransmitters [76], while the presence of oxytocin at levels of 15 ng/mL can support the neurological process by compensating for social information and facilitating the feedback cognition of facial expressions, both happy and sad [75,77]. Therefore, neuroendocrine homeostasis between excitatory (DA, NA, and glutamate) and inhibitory substances (gamma-amino-butyric acid and serotonin (SE)) reflects a sympathetic or parasympathetic predominance, respectively, and acts together with the amygdala and the motor cortex (Figure 3).

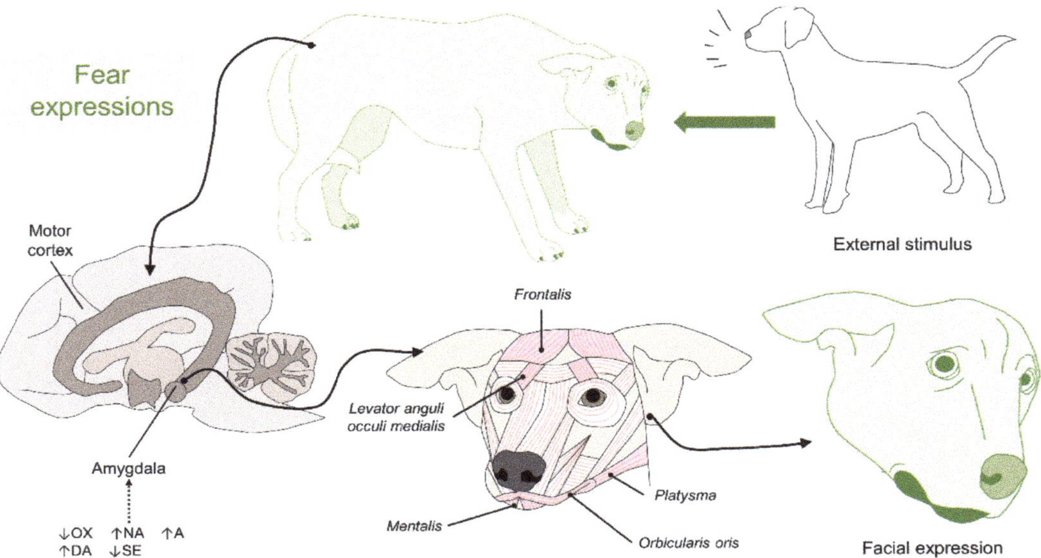

Figure 3. Neurobiology of facial expression in dogs. During a threat exposure, such as another dog, unfamiliar person, or a dispute over territory, the neural mechanism in the amygdala reacts to catecholamine secretion (A and NA). The catecholamines stimulate the motor cortex and its efferent fibers to modify a facial expression. Movements such as flattening the ears to the side (EAD105) and lifting the upper eyelids to have a wider field of vision (AU101) are characteristic of an expression of fear. OXT: oxytocin; NA: noradrenaline; A: adrenaline; DA: dopamine; SE: serotonin.

In the first place, DA is associated with positive or reward-related emotions processed by cortical structures (medial prefrontal cortex, anterior cingulate, and olfactory cortex) and subcortical structures (striatum, amygdala, and hippocampus). These signals are integrated into the ventral tegmental area (VTA) and the *nucleus accumbens* [78].

In this sense, reduced concentrations of DA have been associated with pathologies such as chronic anxiety, while increases are related to events that trigger positive emotions, such as sports and training, together with other catecholamines (NA) and SE (Figure 4) [79]. Likewise, during techniques to reduce stress in dogs (e.g., petting), plasma levels of NA and SE (397.76 pg/mL and 542.75 ng/mL, respectively) have been associated with a relaxed state, a reaction that is also correlated to some lateralization of facial and body language [80]. SE also plays a role in the presentation of aggressiveness and the adjustment of facial expression, where low levels of SE in the prefrontal cortex are present in animals prone to aggressive behaviors [81].

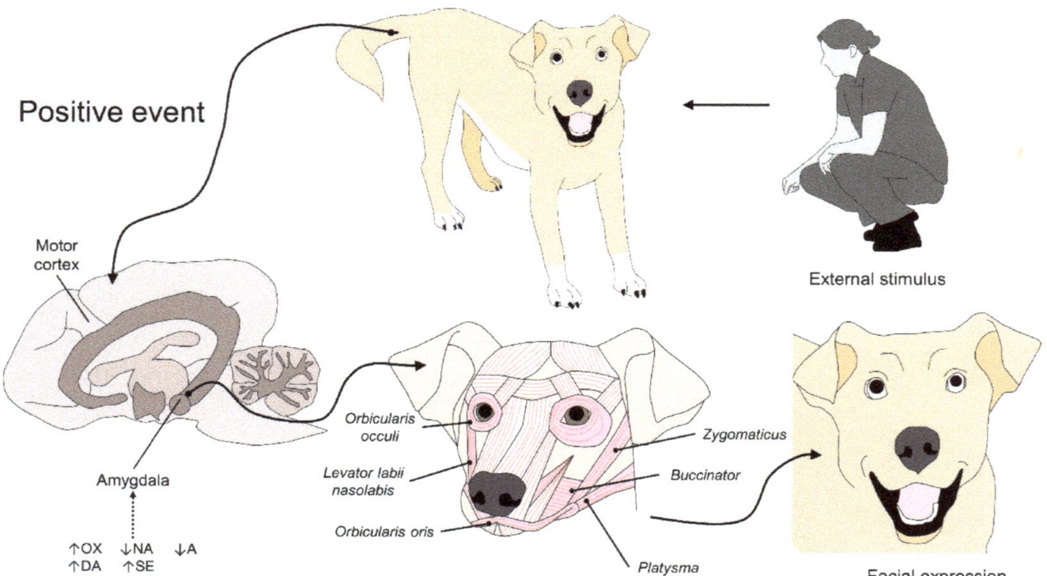

Figure 4. Endocrine and motor control of facial expression during pleasant emotions in canines. When the amygdala associates the presence of humans or conspecifics to a positive emotion, the levels of OXT, DA, and SE increase in response to the stimulus; this activates the facial units that enlarge the eyes and retract the muscles of the mouth to produce a simulated "smile". OX: oxytocin; NA: noradrenaline; A: adrenaline; DA: dopamine; SE: serotonin.

On the other hand, OXT reduces amygdala activation in response to facial expressions and negative emotions like fear or rage by intensifying the connection between it and other regulators, namely the orbitofrontal cortex, cingulate gyrus, and temporal sulcus [82]. Studies in humans show that OXT fosters the capacity to infer the affective mental states of others based on subtle facial signals [83]. Low basal OXT levels have been associated with positive emotions in various species [84]. In dogs, Mitsui et al. [85] determined increased levels of urinary OXT in six Labrador Retriever dogs exposed to positive events such as eating, exercising and stroking ($p < 0.05$, $p < 0.05$, $p < 0.01$, respectively), which reflects relaxation, calm and security. Moreover, research has further ascertained that animals show facial expressions associated with emotions and that low basal oxytocin levels mediate these. In this context, Lansade et al. [86] evaluated the facial expressions of Welsh mares under two treatments of tactile stimulation (grooming): one traditional treatment did not consider the animals' reactions but provoked avoidance, while the other treatment was directed towards regions that the mares considered positive or pleasant. Their work evaluated the horses' facial expressions by observing the positions of the ears, eyelids, lips, and neck, among other gestural indicators. They determined that the cordial, pleasant grooming of the mares significantly increased expressions seen in the position of the neck, which rose to a medium height ($p = 0.01$), accompanied by semi-closed eyes ($p = 0.01$), lips contracted and extended laterally ($p = 0.003$), and the upper lip extended and immobile ($p < 0.0001$). These findings led them to conclude that the facial expressions were modified markedly in the presence of a low concentration of oxytocin compared to basal ones, suggesting that they could be considered positive emotions in horses. Their findings also reveal that oxytocin is an endocrinological marker of emotional well-being responses.

Increases in peripheral OXT after positive human-dog interaction have been shown [87,88], and, concurrently, the administration of exogenous OXT has also been shown to influence how dogs respond to human emotions and during positive interactions. Somppi et al. [89]

evaluated 43 dogs of 16 different breeds. Gazing behavior and the diameter of the pupil were assessed to determine the emotion of the animals. Dogs receiving nasal OXT fixated more often at happy human faces than angry faces and had a larger pupil diameter. The authors attribute this response to the influence of OXT on attention and emotional arousal, which decreases vigilance reactions, diverts attention to positive events, and facilitates human-animal and animal-animal communication, as well as the mediation of social perception and emotional states in canines.

These studies confirm that the generation of facial expressions is due primarily to the decodification of external stimuli by the SNC through connections between the limbic system. Thus, this system is also responsible for responding to facial or non-facial stimuli and even positive or negative emotions, the former mediated by the release of hormones like oxytocin that mediate motor changes in facial muscles. In animals, whose limitation is being unable to self-report their mental state, it is essential to understand the neurobiological mechanisms involved in the perception and development of emotions [78]. However, despite these endocrinological alterations, in dogs, the question remains whether facial expression reflects emotions and serves as a communication tool or is the result of the manipulation and mimic of human gestures [90].

6. Emotions in the Dog

Anderson and Adolphs [91] define emotions as internal states of the CNS that generate physiological, behavioral, cognitive, and subjective responses to manifest distinct categories of states. In humans, emotions constitute a complex behavioral phenomenon involving multiple levels of neuronal and chemical integration [92]. These internal states result from the evolution of primary emotions (anger, fear, disgust, happiness, sadness, and surprise) and are associated with well-defined neuronal and neurophysiological chemicals that act as supporting factors to ensure the performance of specialized survival functions [35,93].

In general, gestures communicate emotional states and allow individuals to appraise social interactions [94]. Therefore, facial expression is an effective non-verbal information tool [95] to promote, create and facilitate social interactions between animals [96]. Consequently, facial expressions are one of the principal elements that allow individuals to identify the emotions expressed by others in a social group [19,97] and are the main pathway for transmitting the affective information that leads them to act in any given situation involving an emitter and a receiver of facial signals [98]. Even though it is still controversial to state that dogs use facial gestures to communicate their mental state, in a study conducted by Karl et al. [99], the response of 12 domestic dogs to positive social and non-social neutral stimuli was assessed using functional magnetic resonance imaging. The results showed that limbic areas such as the left amygdala, hypothalamus, and insula were activated during positive social interactions. The above shows the correlation between the neural response for the processing of emotions and the possible facial response, which would support the theory that dogs can transmit their emotional state using facial expressions. Besides, the personality and empathy of people have been correlated with their ability to interpret the emotions of animals. In this sense, it has been reported that empathic people can discriminate between positive and negative affective states [100]. However, independent of human perception, the possibility that dogs can transmit their emotional state with facial expressions cannot be denied.

Siniscalchi et al. [40] mention that domestic dogs use a broad range of facial regions to communicate their emotional condition; however, the orbital region plays a crucial role in interacting with conspecifics. For example, in stressful situations, the dog opens its eyes, exposes the white sclera, and flattens the ears (EAD105), gestures that indicate warning of an imminent threat [41,101]. In this context, Caeiro et al. [102] evaluated the response of the face of dogs and humans during the perception of different visual and auditory stimuli using the FACS system in both species. The authors observed that dogs emit facial movements depending on the perceived stimulus. In addition, they observed that the facial movements are different from those observed in humans during similar emotional states. Similar facial movements have been identified in dogs and humans, such as the eyebrow-raise or the upper lip raise (AU110). However, it may be inappropriate to consider that they have the same emotional valence. For example, a raised upper lip is often mistakenly interpreted as a smiling expression by children [103].

Similarly, Konok et al. [5] reported that humans attempting to describe dogs' emotions have an anthropomorphic influence. Although relevant for animal welfare, humans tend to interpret the emotional states of animals in the same terms they use to perceive their own. However, as stated before, it is not appropriate to assume animal emotions based solely on human perception and anthropomorphization of human feelings or facial expressions. On the other hand, Bloom et al. [104] reported that 105 liberal arts students were able to successfully determine the emotional states of canines (joy, sadness, surprise, disgust, anger, fear, and neutral) using the DogFACS, which has proven to be an adequate tool to assess emotions and facial expressions [25] (Figure 5).

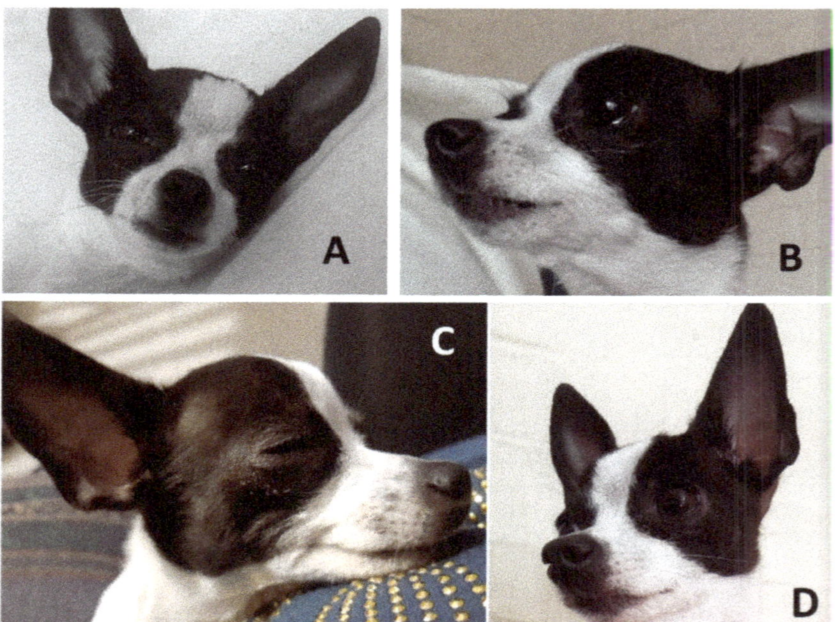

Figure 5. Facial expressions in dogs and their links to humans. Image of a one-year-old female Chihuahua showing marked changes in facial expressions upon recognizing similar emotions in her owner's gestures. (**A**) Neutral. Relaxed expression; (**B**) Sadness. Broad forehead, large eyes, ears erect with caudal inclination, closed snout; (**C**) Pleasure. Tension in the labial commissure, eyes semi-closed with facial mimicking that simulates the form of a human smile; (**D**) Surprise. Eyebrows raised to show apparently larger, more expressive eyes, with ears raised and labial tension.

Facial movements are also considered an adaptive response to threatening stimuli. However, their interpretation depends not only on conspecifics but also on the humans with whom dogs establish bonds [105]. Flint et al. [106] have evaluated the association between facial and body movements, a negative emotional response such as fear, and the reliability of the owner to identify these changes. Through videos, the accuracy of the owners of 735 dogs was evaluated to detect fear-related behaviors, focusing on the head, changes in posture, and facial and body language. The authors found that owners accurately identified the position of the ears (flattened ears (EAD015)) as a facial modification associated with fear with a sensitivity and specificity of 0.76 and 0.88, respectively. Other changes reported with a sensitivity and specificity above 0.75 were a low posture, lowered tail, hiding, escape attempts, panting, yawning, lip licking, and no eye contact. Additionally, based on the behaviors and postures identified by the owners, a training tool was developed to improve the recognition of fear behaviors. Although the differences after the training did not significantly influence the reliability of assessing fear in dogs, owners were more likely to recognize "mild to moderate fear" (OR: 1.01, 95% CI: 0.86, $p = 0.881$) based on the mentioned facial expressions such as the position of the ears and the other related behaviors. In contrast, as stated by Bremhorst et al. [47], when exposed to positive events (food and toys), 29 dogs of one breed showed erect ears (AED101) and blinking (AU143 + AU145). Under negative conditions (ignored by the owner), nose licking, drooping jaw (AU26), drooping lip (AU25), and flattened ears (EAD105) were predominant (Figure 6). Therefore, both studies show that recognizing these facial movements, particularly the position of the ears, could contribute significantly to the identification and interpretation of dogs' emotional state, representing a wide field to study in the future.

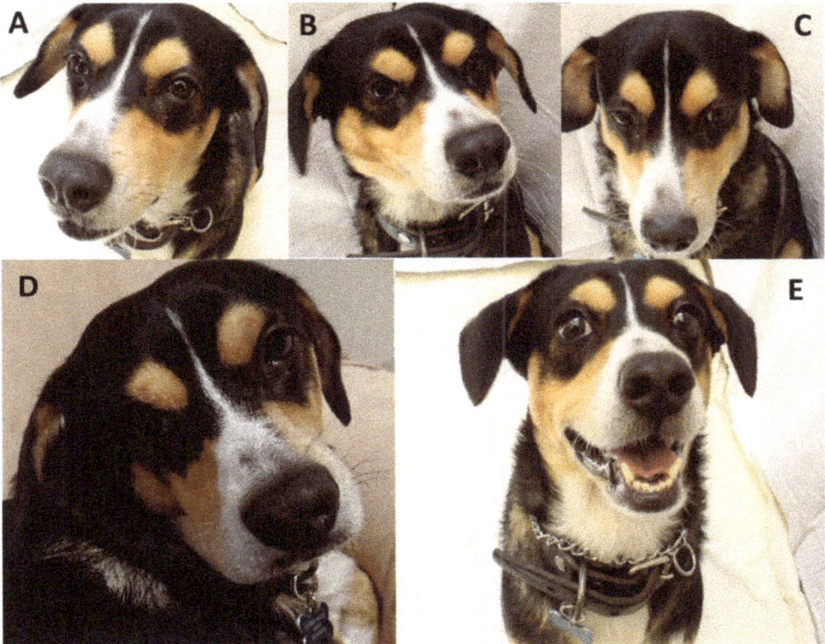

Figure 6. Changes in facial expression during exposure to various stimuli. (**A**). Neutral. No relation or expression; (**B**). Anger. Observations include a fixed gaze with ears at medium height or pulled back with flared nostrils and tensed cheeks; (**C**). Fear. The gaze has deviated with the ears pulled back, exposing the frontal region as a signal of submission; (**D**). Surprise. The eyebrows are inexpressive, perhaps with a slight raising of the medial section to give the appearance of large eyes, with the ears at medium height; (**E**). Pleasure. Eyebrows and ears are clearly raised with the mouth open and tension in the labial commissure.

Since dogs are highly sociable animals with complex groups, understanding heterospecific emotions is of particular importance [107]. Racca et al. [108] demonstrated this in their study that assessed dogs' capacity to discern facial changes based on visual stimuli and if such behavioral responses are influenced by the species observed (i.e., human vs. conspecific faces). Their study protocol involved seven dogs of different breeds in two phases. The first type of discrimination was determined using images of (i) a familiar visual stimulus and (ii) a novel one. They found that the dogs showed greater interest in the novel stimulus and could discriminate objects presented simultaneously. In the second stage, a simultaneous face discrimination test using images of humans vs. different breeds of dogs, the authors found that the time the dogs invested in observing the images averaged 4.2 s more for the new rostrums of both humans and conspecifics. Those authors concluded that canines could perform face discrimination regardless of species and modify their facial expression responses depending on the type of face shown. In a posterior study by the same authors, the ability of dogs to process facial expressions was compared to those of 4-year-old children, showing them a series of photos with emotional facial valence (positive, negative, or neutral). In this case, they videotaped the subjects' facial expressions and eye movements. The authors found that the dogs spent less time observing the images (3.87 s vs. 4.51 s in children) and reported greater eye movement latency (1.21 s vs. 0.66 s in children). The comparison of the responses of the two groups detected a significant difference in the processing of the positive images, as opposite responses occurred concerning these emotions; one example of this was deviating the gaze, which was identified as the behavior that canines exhibit when distinguishing between happy and sad faces [109]. Both results confirm sensitivity to facial expressions such that dogs can interpret and react despite being individuals of different species, as in, for example, the gaze response to a threatening expression (Figure 7) [110]. A particular communication mechanism between dogs and humans has been observed, as stated by Ogura et al. [111]. The gazing behavior was evaluated in seven dogs of different breeds exposed to different situations: humans using hand signals, other dogs, and the presence of a cat. The dog paid more attention to the hand signals of humans but focused more on the face and body when looking at conspecifics and cats. This research raises a probable species-dependent communication difference.

For these reasons, facial expressions that reflect different emotional states may well prove to be an effective tool for (i) achieving non-verbal communication [95]; (ii) creating, fomenting, and facilitating social interaction among animals; (iii) constructing or improving successful social groups, as mentioned by Brudzynski [96], and the affiliative function of expressions that promotes contact and closeness of members to each other [84]; and (iv) providing relevant information that, together with other gestural signs, may modify the interpretation and evaluation of such communications [19,97]. In this regard, fear and anxiety refer to what we consider emotional responses to aversive stimuli. These emotions are adaptive for survival because they are triggered to avoid a perceived or anticipated threatening stimulus [105]. It has been established that those behavioral problems in dogs are probably associated with a close relationship to humans. Hence it is vital to avoid these states [112], as Mariti et al. demonstrated [107]. Those researchers further concluded that the dog owners did not require experience to identify these negative emotions accurately. Owners were able to identify the behaviors and expressions associated with aversive states without needing any training to identify them. This will be discussed in detail later.

In addition, the ability of dogs to read the gestures emitted by humans may be related to another phenomenon that occurs when individuals from social groups recognize certain expressions from the same species or others and respond with a similar or antagonistic pattern [104]. This process is called "emotional contagion." The above has been reported in rats, animals that, after encountering conspecifics with pain, manifested pain gestures and behaviors without having any injury [113,114]. This suggests that animals can display a similar state after reading the facial language, although facial mimicry does not reflect the meaning of the inherent emotion on every occasion [115]. Using emotional contagion, researchers have sought to ascertain whether dogs can show interest in the emotions expressed by other individuals in their social circle. In this field, Katayama et al. [22] observed that the emotional contagion of humans towards dogs occurs more pronouncedly with women and is affected by the surroundings in which it occurs. They evaluated the response of 34 dogs towards their owners while they carried out a reading test with an audience, considered a stressful event for humans. They reported a correlation between the R intervals of the heartbeats present in dogs and humans ($r^2 = 0.74$, $p < 0.01$), and that in addition, the emotional contagion of humans towards the dog was influenced by the sex characteristics of the pet and the time of coexistence, since in females and in individuals who had a long time of coexistence with the owner, the emotional contagion was higher.

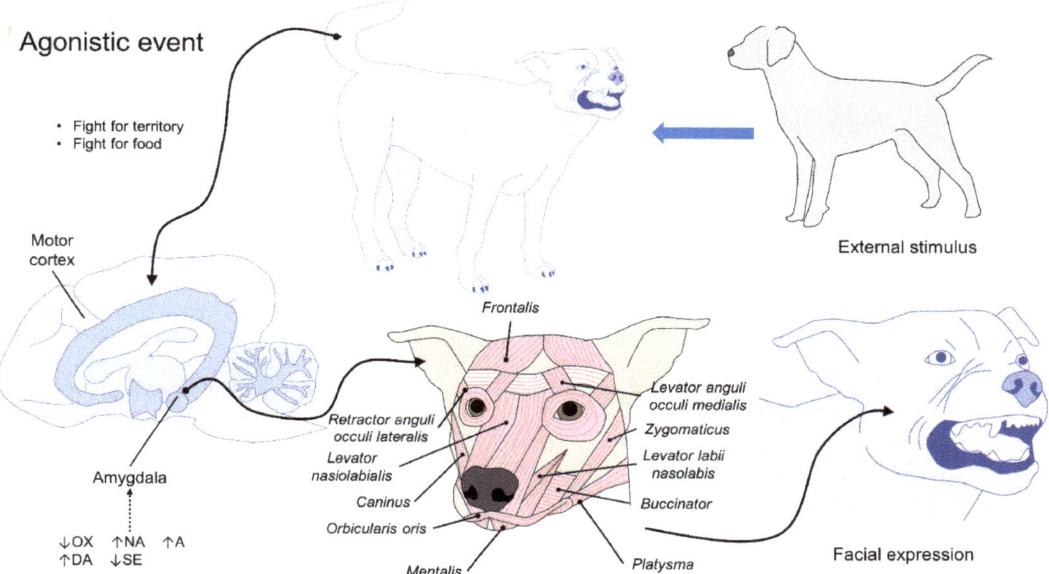

Figure 7. Facial expressions represent aggression. The perception of an adverse external stimulus (e.g., a person or another animal) triggers a neuroendocrine response with an elevation in catecholamine levels, together with a decrease in circulating levels of OXT, DA, and SE. When this negative emotion is recognized in the amygdala, its connections to the motor cortex and facial muscles (trough the facial nerve) cause the flattening of the ears (EAD105), lifting of the upper lip to bare their teeth (AU116), nostril opening (AD38), more visible white sclera (AU101), and vocalizations (AU34) as a sign of threat. OXT: oxytocin; NA: noradrenaline; A: adrenaline; DA: dopamine; SE: serotonin.

Although this phenomenon is a primitive form of empathy between individuals who share the same social environment, it does not necessarily involve a higher psychological function [23]. However, it is a process that improves the human-animal interaction, particularly in domestic canines who have adapted and refined mechanisms to live inside mixed-species groups [98].

As can be seen, comprehending the emotions and affects of heterospecifics is of great significance for domestic species, such as dogs [110]. Recognition of emotions, mimicry, and gestures is considered of greater importance than other forms of expression [115], though facial mimicry does not necessarily accurately reflect the meaning of the underlying emotion [112]. This phenomenon has also been identified in dogs because their ability to recognize certain expressions shown by individuals in their social circle—whether of the same or a different species, including humans—allows them to generate similar or opposing responses to those manifestations. Those responses take the form of facial expressions that reflect perceived states [102] (Figure 6).

Various emotional and behavioral problems described in dogs are attributed to their close interaction with humans and modern lifestyles [116]. These led Kurachi et al. [117] to state that behavioral problems in animals must consider their physical and emotional origins and the human-animal bond that can influence their welfare. To achieve an acceptable level of welfare, animals require positive events (e.g., pleasure) instead of only preventing negative ones (e.g., fear or anxiety) [4,118].

In this context, further research is required in dogs to evaluate positive emotional states and associate them with facial and body movements, as well as in other species, like in pigs [119], in sheep [120], and in horses [121], among others. The emotional state influences the behavior, communication, social bond, and cognitive functions of the group's individuals [122]. Then, the emotions are expressions of the internal states of the CNS reflected in animals' physiological and behavioral responses, which foster better interaction among the individuals that make up a social circle. Identifying these expressions is an additional, important indicator for recognizing the degree of welfare that animals experience.

7. Interpretation of Perception and Expression Interspecies

Currently, there is a growing interest in interpreting dogs' emotions to enhance the welfare of both species: dogs and humans [39,123]. Flint et al. [106] asked a group of dog owners to distinguish expressions of fear in their pets to identify animal facial expressions linked to aversive stimuli. Using videos, they demonstrated the importance of several regions, especially the head, body posture, and facial expressions. The owners adequately recognized low posture, ears pulled back and tail between the legs or curled downwards as behavioral gestures that indicated concealment, attempts to flee, panting, yawning, lip-licking, and null visual contact. In this context, Bloom and Friedman [25] designed a study to ascertain people's ability to identify facial expressions emitted by dogs in the presence of diverse stimuli and, on that basis, attempted to define such emotions as happiness, sadness, surprise, disgust, anger, and fear. They utilized positive and negative stimuli to provoke dogs, and they found that when their owners emitted positive stimuli, the dogs' reactions included wide-open eyes, erect ears, and a fixed, forward gaze (Figure 6E). When the element of surprise was added, the dogs showed furrowing in the central region between their ears, a slightly inclined head, raised eyebrows, and the mouth closed or only slightly open. In contrast, when negative stimuli were shown, the dogs tended to bow the head and deviate their gaze laterally, accompanied by a differential raising of the two eyebrows, caudally inclined, erect ears, and the mouth closed or slightly open (Figure 6C). When given medication, the dogs inclined their ears towards the caudal region (EAD105), bared their teeth slightly (AU118), and showed a slight lowering of the nose (AD40) and eyebrows (Figure 6C).

When dogs are induced to feel fear through verbal communication, their ears are markedly erect; their eyes show a small display of the sclera, their gaze shifts from side to side, and the mouth is either slightly open or tightly shut (Figure 8A). Finally, upon hearing the negative stimulus, bad boy, "the dogs exponentially bared their teeth and fixed their gaze on the objective with eyebrows pointing towards the center, ears folded rearwards, and the mouth semi-open or open (Figure 8B).

In the same study, the expressions of the animals in the absence of a stimulus (neutral) were also evaluated. The resulting photos were like those taken after a positive stimulus—"good boy—, with a slight lateral inclination of the ears. In contrast, Racca et al. [109] determined that, in dogs, a neutral human facial expression is categorized as a negative emotion. They concluded that the human recognition and interpretation of dogs' facial expressions could trigger a certain degree of expected behavior by the owner or handler and improve their interspecies interaction.

It is important to emphasize that no previous experience or training is necessary for people to identify emotions in dogs based on their facial expressions. Comparative studies have shown that the ability to interpret animal behaviors is influenced by the level of experience of the evaluator [124]. Lakestani et al. [125] compared the ability of children and adults to understand the friendly and aggressive behavior of dogs. In this study, young individuals could not discern friendly from aggressive behaviors ($p < 0.001$). This was attributed to children focusing more often on the face of the animal, leading to misinterpretation ($x^2 = 80.2$, df = 1, $p < 0.001$). These results reinforce the idea that the experience and ability of the evaluator influence their ability to decipher facial expressions and body language of companion animals. However, a study by Tami and Gallagher [123], which compared observations of the body language and facial expressions of untrained people to those of individuals who had some knowledge of canine behavior (e.g., dog trainers or owners), demonstrated that both groups of evaluators found it challenging to distinguish between aggression and playfighting. For this reason, integrating facial expressions with other gestural signs could be a key to elaborating a comprehensive assessment of dogs in this regard.

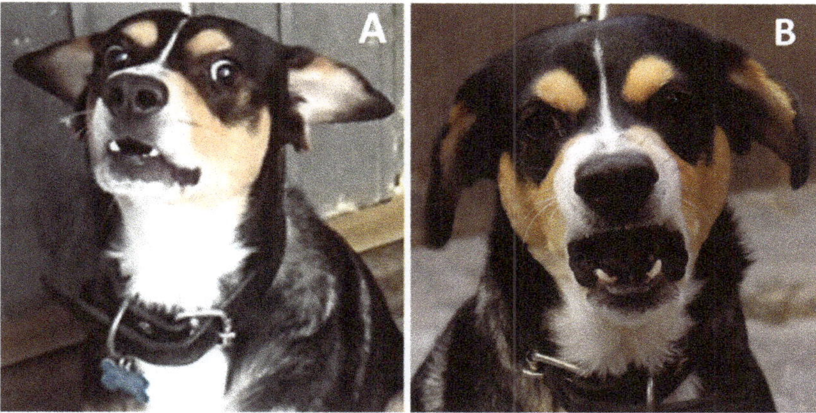

Figure 8. Facial expressions related to fear can provoke aggression. (**A**) Nose, sclera, and snout. Increased exposure of the sclera is caused by raised eyebrows and tension in the nasal region with flared nostrils, accompanied by an open mouth, raised upper lip, and exposed fangs. Additionally, dogs may emit vocalizations. (**B**) Ears and gaze. The positioning of the ears is commonly seen to be at a medium height concerning the frontal region; the animal's gaze is fixed in the direction of the threatening stimulus.

Humans perceive raised eyebrows, characterized by an increase in the overall size of the orbital region, as a sign of sadness [21]. According to the description by Merola et al. [126], a backwards gaze seems to indicate a request for some object and to verify humans' reaction to an ambiguous object. In this sense, people also confer an anthropomorphic sense to the gaze of animals, particularly the so-called "guilty look," which is associated with this feeling of guilt by around 74% of people when observing said facial expression. This gesture is characterized by ethologists by an aversive gaze, lowered ears, constant licking, and a lowered head position. However, there is a difference between what humans interpret and the actual reason for the facial motor response. For example, in a study with 14 mongrels and purebred dogs, there was no difference in the prevalence of guilty gaze when the animals performed an acceptable behavior or not. The authors concluded that dogs respond with a guilty look when scolded, regardless of having infringed an instruction or not, which means that the animal does not understand that it has disobeyed and lacks a cognitive association [127].

In recent years, eye-tracking systems have become relevant to distinguish facial language associated with positive or negative stimuli [128]. In these studies, canines confer more attention to body language, unlike humans, who prioritize the grimaces of fellow humans and other animals. Correia-Caeiro et al. [129] evaluated the reaction of 129 humans and 92 dogs to videos with gestures of positive (happiness, pleasure) and negative (anger) emotions from both species. The included animals responded with several gestures to expressions of conspecifics and humans, with no evidence of facial mimicry towards the recordings. Interestingly, the action of the EAD102 (ear adductor) unit was more frequent when observing other dogs' faces, which can be interpreted as a positive response to the same species [130,131].

Another element of facial expressions that needs evaluation in dogs involves the lips, since these function as messengers to other individuals that express aggressive intentions; that is, displays with lips retracted, the snout thrust forward (simulating smaller lips), with the mouth open. These expressions reflect a threat of a particular type or degree [132] (Figure 6B). Comparatively, between humans and dogs, Action Unit 27 (AU27), representing the wide opening of the mouth, and AU6, where the muscle around the eyes contract and the cheeks are raised as a "happy face," are similar facial expressions in both species [129]. Similarly, lip-licking in dogs is understood as a behavior involved in identifying expressions of anger in humans [98].

Firnkes et al. [133], in addition, reported that a deviated gaze and lip-licking are associated with appeasement. Appeasement includes all those behaviors aimed to avoid confrontations with conspecifics and other species. The objective of their research was to examine manifestations of these signals towards unfamiliar people using a standardized behavioral test. Their findings suggest that a deviated gaze and lip-licking occur during appeasement and, more often, when dogs face situations involving threat or conflict. However, these signals decrease when the threat is greater and are replaced by behaviors indicative of submission (Figure 6C) or attempts to flee, defining submission as those behaviors used to avoid the threat represented by a dominant member of the same species. Those authors concluded that dogs seem to have the ability to employ distinct strategies according to the type of situation. Additionally, they were able to identify that lip-licking forms part of dogs' greeting behavior towards humans, since socio-positive focuses were observed with greater frequency. For this reason, this gesticulation can play a crucial role in intra-specific communication by serving as a signal that expresses peaceful intentions.

Regarding evidence for the lateralization of facial movements in domestic canines, Nagasawa et al. [134] analyzed whether dogs could do this when encountering their owners. Lateralization of emotion is related to activating the right or left hemisphere following positive or negative emotion recognition in the brain. The goal was to determine whether lateralization was correlated to positive and non-positive social events (presence of the owner, or toys, respectively). The authors found that when the stimuli were of a social type, dogs showed greater laterality towards emotive stimuli by responding with a facial expression involving the eyebrows or ears of the left side, compared to the case of inanimate objects that they had to avoid, where they showed greater facial expression on the right side. Similarly, lateralization of the gaze has been reported and related to the ability of dogs to recognize emotions and activate the right or left cerebral hemisphere, a relevant characteristic in animals with frontal eyes, whose visual connection usually responds to activation of the contralateral hemisphere. For example, a gaze to the left indicates activity in the right hemisphere, associated with neutral human faces. Racca et al. [109] evaluated the reaction of 21 canines of various breeds to negative, neutral, and positive expressions of conspecifics. The results showed that the maintenance of a gaze to the left, in response to a negative expression, was greater (1.60 ± 0.20 s) compared to the gaze to the right (0.57 ± 0.09 s) and neutral and positive stimuli (1.26 ± 0.16 and 0.65 ± 0.1[illegible] s, respectively). This can be translated as a predominance of right hemisphere activity, as the right hemisphere is a structure involved in interpreting negative emotions. Moreover, a prevalence of head-turning to the left in animals exposed to anger or fear human gestures has been reported [132].

It is worth mentioning that there are current reports that define limitations in the assessment of the emotions through the eyebrows and lips in the dog. Some examples are a paedomorphic face, the type, length, and color of the hair, and the anatomical differences between breeds (e.g., the structure of the skull) [104]. Additionally, dark-colored spots on the face of canines, particularly in the eyebrows, tend to give an aggressive appearance and are considered another factor that can alter the perception of humans regarding the emotion of companion animals [132,135].

As can be seen, two critical regions to be considered when evaluating facial expressions in dogs are the position of the lips and eyebrows, since they allow us to distinguish between stimuli perceived as positive vs. negative. Dogs capture these stimuli to identify intentions such as threat or play, though they present a certain degree of difficulty for assessment because some may be similar and impede the ability to arrive at accurate interpretations of the emotion shown.

8. Changes in Facial Expressions Related to Pain

The identification of facial expressions in animals offers an opportunity to assess, in a non-invasive way, the welfare of domestic species [16]. As stated by the International Association for the Study of Pain, pain is "an unpleasant sensory and emotional experience associated with, or resembling that associated with, actual or potential tissue damage" and has a connection with gestures and the degree of pain [136]. The current approach to pain assessment is based on the affective magnitude (e.g., "how does it make you feel?"). Therefore, facial expressions are an alternative to evaluate pain and differentiate this state from other emotional states such as fear, stress or anxiety (Figure 9) [26,137,138].

With the implementation of FACS, it has been possible to identify subjective states such as pain. The tension of the muscles of the face, lips, orbital region, and flattening of the ears are commonly recognized in animals [2,139–150]. This system has been adapted as a veterinary clinical tool (called Grimace Scales) to identify pain in species such as equines [142], chimpanzees [60], macaques [143], sheep [8], rabbits [144], ferrets [145] and rats [33].

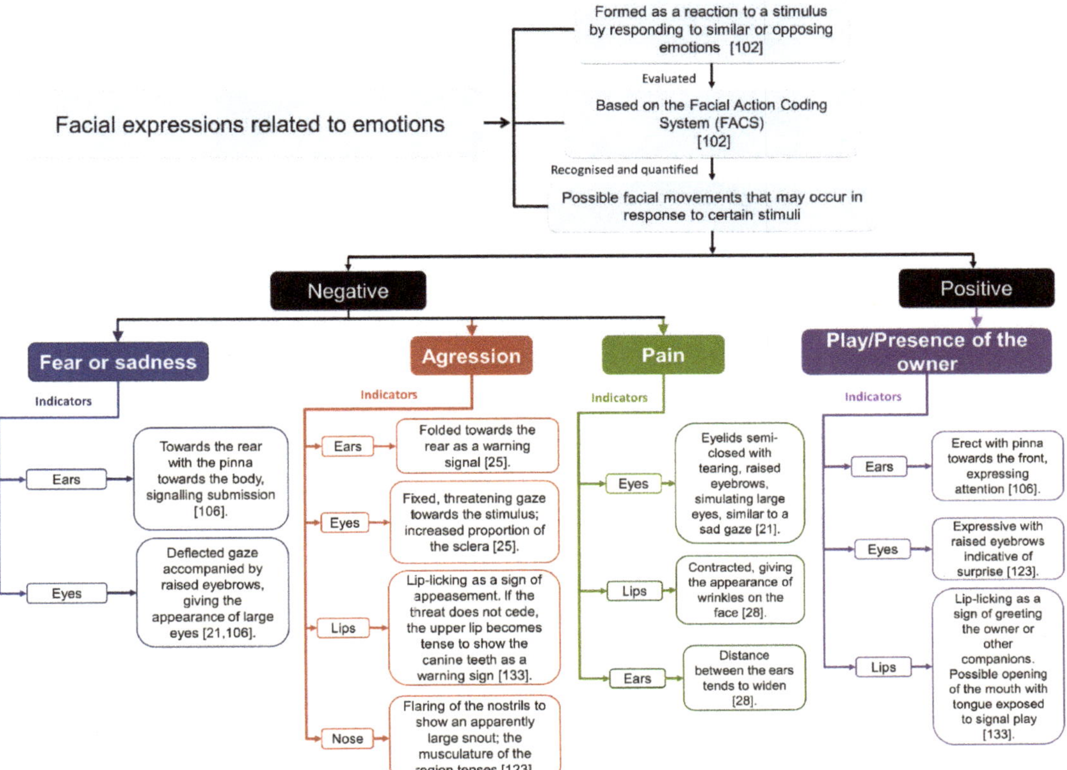

Figure 9. Facial expressions related to emotions [21,25,28,102,106,123,133].

Behavioral evaluations of pain must consider several indicators. In dogs, Camps et al. [146] described the principal characteristics of aggression related to pain. They affirm that behavioral changes are the most common indicator of pain manifested by this species through such behaviors as aggression, fear, vocalizations, reduced interaction with conspecifics or family members, and altered postures or facial expressions, restlessness, and concealment (Figure 10).

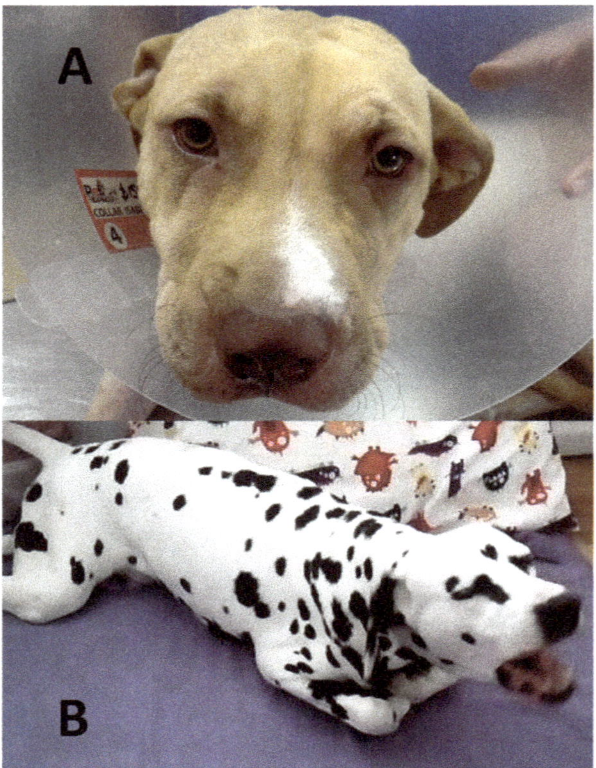

Figure 10. Expressions related to pain. (**A**). Mild pain. Expressions are characterized by a sustained raising of the eyebrows with displays of the frontal region of the head and lateral abduction of the ears, generally associated with a sad face, as represented in this image of a one-year-old female Pitbull, which shows a slight anaphylactic reaction. (**B**). Severe pain. Expressions are characterized by semi-closed eyelids, cheek tension, and vocalizations, as can be seen in the image of this two-year-old female Dalmatian with acute polyradiculoneuritis.

In equines, facial characteristics such as asymmetric ears, the tension of orbital muscles, and the lips decrease their frequency of appearance during the administration of analgesics [147]; these characteristics are known as the "face of pain" [140]. On the other hand, in felines, observed facial movements positively correlated with the Feline Grimace Scale (r^2 = 0.86), a proposed validated tool to evaluate nociception in cats [29].

In felines, the Feline Grimace Scale (FGS) has been implemented to identify acute pain using facial expressions, with an efficacy of up to 94% to differentiate between pain and pain-free cats. These scales have made it possible to identify that 85% of the facial changes observed during a painful event are concentrated in the ears, eyes, whiskers, and nose [30]. As previously mentioned, the movements of these regions serve as clinical assistance for the identification of pain. The evaluation of these changes with the FGS has reported overall inter-rater reliability of (ICC) of 0.89 (95% CI) and intra-rater reliability of 0.91 [31], which denotes the clinical utility of facial expressions. In that regard, facial expressions must be considered when evaluating acute pain in dogs. Pain scales, such as DIVAS (Dynamic Interactive Visual Analogue Scale) and the pain evaluation scales developed by the University of Melbourne (UMPS), the University of Colorado (CSU-CAPS), and the Composed Measure by Glasgow (CMPS), include changes in facial expressions as a part of their assessment. A fixed gaze (Figure 6A), squinting (Figure 6B), or a grimace with a fearful look are examples of this. They also incorporate a glassy appearance of the eyes, mydriasis, furrowed brows, lips drawn back, and flattening of the ears [39,141].

DIVAS includes evaluating changes in facial expression and behaviors and vocalizations derived from pain, such as changes in posture, inactivity, or aggressiveness. A scale from 0 to 100 mm has been used to assess the effects of preemptive analgesia in routine surgeries such as ovariohysterectomy; a patient with a score of 40 mm requires analgesic intervention [148]. Likewise, it has been used to estimate pain intensity in canines undergoing various surgical procedures, such as exploratory laparotomy, hemilaminectomies, osteotomies, and atlantoaxial subluxation [149]. On the other hand, the UMPS considers six categories in which the animal's appearance includes observations such as the ears' position and the form of the eyes to classify the changes derived from moderate to severe pain [150].

Finally, the CSU-CAPS considers facial changes such as droopy ears, arched eyebrows, and darting eyes (known as a "worried facial expression") to assess the level of pain from 0 to 4 (minimal to severe pain), considering a value greater than 2 as a rescue point [151]. Likewise, these parameters have also been proposed as an evaluation method in cats. An example of this is the validated Multidimensional Composite Pain Scale of the University of Botucatu (UNESP-MCPS). In this scale, psychomotor and facial aspects are considered a means of manifestation of pain through changes in facial expression. These include assessing the distance between the ears and the facial muscles' tension as distinctive features of pain recognition with a sensitivity of 74.6% and specificity of 84.6% [152,153]. This has been used to evaluate postoperative analgesia in procedures such as tibial plateau leveling osteotomies [151] or to determine the perioperative efficacy of opioid drugs such as tramadol during canine sterilization [154].

Nevertheless, all these approaches do not consider specific regions or the distance required to assess the facial expressions and correlate them to the degree of perceived pain. In contrast, in cats, the CMPS-Feline is a scale in which facial evaluation has been included together with behavioral modifications to evaluate postsurgical pain. On a scale from 0 to 20, where 5 or more is considered the analgesic intervention point, facial movements such as the muzzle shape and the ears' position have been shown to correctly identify 78.6% of animals that required analgesia [155].

For dogs, to date, there are no reports that can associate the intensity of pain to specific facial traits. Therefore, changes in the lips, eyes, and facial muscles are valuable characteristics that can be integrated into the existing pain scales to help identify the intensity and degree of pain in dogs, as done in other species.

Animals cannot communicate pain in a verbal way. Therefore, the study of facial expressions in animals can aid in identifying and differentiating between emotional states such as fear of pain. However, some limitations are present in the implementation of facial changes in pain scales. For example, scales do not consider the residual effects of anesthetics or the sedation level caused by some local analgesics, opioids, sedatives, and cyclohexamines [156]. Additionally, it is important to mention that the accuracy of these scales also depends on the type of pain (i.e., most of the scales have been designed to assess postsurgical pain) [157]. The age, experience, training of the evaluator, and visual acuity when attempting to accurately quantify the degree of pain experienced by a dog based on observations of facial, posture, or behavioral changes can also influence their interpretation [149]. In comparing veterinary students and anesthesiologists, the former gave high scores to those animals that the anesthesiologists rated as low; the study concluded that there was 95% of no agreement between both evaluators [149]. Likewise, the scales are a subjective method that can be vulnerable to evaluator errors or biases. Hence, they are considered an additional tool for the study of pain, since there is no single biomarker, physiological parameters, posture, or facial expression that, by itself, can be considered as the "gold standard" of pain assessment [155].

Further studies in dogs also require the development and implementation of techniques and tools currently applied in other species, such as artificial intelligence and computer vision systems for the recognition of pain gestures in humans and equines [158], or the automatic detection of Facial Units in sheep, a technique with up to 67% accuracy [52]. Likewise, cameras or visual sensors, global positioning systems (GPS), electrocardiography, electroencephalography [159], and infrared thermography [160,161] are alternatives to evaluate emotions and motor responses in dogs comprehensively. It is essential to consider that human perception of emotions can be subjective and that the anatomical characteristics of the dog [162] (e.g., the coat color, the shape of the ears, or the breed itself) can influence and affect the objectivity of facial expression recognition and must be discussed and analyzed before associating facial gestures with a particular emotion.

Today, small species play a particularly important role in the lives of many people by generating a human–animal bond characterized by an acceptance and treatment similar to that of family members [163]. Regarding this, the anatomical differences between breeds represent a challenge for facial expression recognition in dogs. For example, the cranial structural features of prognathic dogs with short muzzles and deep facial folds do not allow adequate recognition of muscle movements and could interfere with the assessment of facial expression in these animals [104]. In contrast, in the case of primitive dogs such as the Siberian or Alaskan Husky breeds, the lack of some facial muscles to perform specific facial movements has been reported [62]. These characteristics can lead to a misinterpretation of negative emotional states such as fear and pain.

Likewise, Bloom et al. [104] have observed that another breed characteristic that intervenes in evaluating the facial expression is the color of the coat. In dogs with black fur, identifying the tension or elevation of certain muscle groups (e.g., supra- or infraorbital) could be a limiting factor. Similarly, small dark spots of hair on the face or in the supraorbital regions lead to a misinterpretation of the meaning of a facial expression [40].

9. Conclusions

Facial expressions in animals, including dogs, are a communication form of positive and negative emotions, in which motor responses are developed in specific facial action units. These units include the eyebrows, eyes, position of the ears, opening of the mouth, and others. These movements are the result of neural circuits whose primary structures are found in the limbic system, the amygdala, the motor cortex, and its derivations towards the facial nerve, the main nerve responsible for innervating the muscles of the face and generating the different expressions based on the valence of the stimulus. These expressions are fundamental to maintaining social interaction between dogs and other species, as in relationships with humans.

Although there is evidence of certain patterns of facial changes associated with specific stimuli in domestic dogs, it is challenging to objectively evaluate facial expressions due to domestication, the close human-animal relationship, and the extensive inter-species variability.

Further research in the development of canine facial action scales and the implementation of tools such as computer vision systems, pain scales, infrared thermography, and others are alternatives to evaluate and recognize the association between facial expressions, emotion, and well-being in pets, as well as to use facial expressions as a clinical tool to evaluate pain and diverse emotions in dogs.

Author Contributions: Conceptualization, D.M.-R., M.M.-R., A.G., A.O., C.M.; investigation, D.M.-R., M.M.-R., C.M., A.O., I.H.-Á., P.M.-M., A.C., A.D., B.R., A.G.; writing—review and editing, D.M.-R., I.H.-Á., M.M.-R., C.M., J.M.-B., A.C., A.G.; project administration, D.M.-R., I.H.-Á., A.G., C.M., J.M.-B., P.M.-M. All authors have read and agreed to the published version of the manuscript.

Funding: This research received no external funding.

Institutional Review Board Statement: Not applicable.

Data Availability Statement: Not applicable.

Conflicts of Interest: The authors declare no conflict of interest.

References

1. Crivelli, C.; Fridlund, A.J. Facial displays are tools for social influence. *Trends Cogn. Sci.* **2018**, *22*, 388–399. [CrossRef]
2. Mota-Rojas, D.; Olmos-Hernández, A.; Verduzco-Mendoza, A.; Hernández, E.; Martínez-Burnes, J.; Whittaker, A. The utility of grimace escales for practical pain assessment in laboratory animals. *Animals* **2020**, *10*, 1838. [CrossRef]
3. Darwin, C. *The Expressions of the Emotions in Man and Animals*; William Clowes and Sons: London, UK, 1872; pp. 1–399.
4. Boissy, A.; Manteuffel, G.; Jensen, M.B.; Moe, R.O.; Spruijt, B.; Keeling, L.J.; Winckler, C.; Forkman, B.; Dimitrov, I.; Langbein, J.; et al. Assessment of positive emotions in animals to improve their welfare. *Physiol. Behav.* **2007**, *92*, 375–397. [CrossRef]
5. Konok, V.; Nagy, K.; Miklósi, Á. How do humans represent the emotions of dogs? The resemblance between the human representation of the canine and the human affective space. *Appl. Anim. Behav. Sci.* **2015**, *162*, 37–46. [CrossRef]
6. Leliveld, L.M.C.; Langbein, J.; Puppe, B. The emergence of emotional lateralization: Evidence in non-human vertebrates and implications for farm animals. *Appl. Anim. Behav. Sci.* **2013**, *145*, 1–14. [CrossRef]
7. Bennett, V.; Gourkow, N.; Mills, D.S. Facial correlates of emotional behaviour in the domestic cat (*Felis catus*). *Behav. Proces.* **2017**, *141*, 342–350. [CrossRef]
8. Wathan, J.; Proops, L.; Grounds, K.; McComb, K. Horses discriminate between facial expressions of conspecifics. *Sci. Rep.* **2016**, *6*, 38322. [CrossRef]
9. Wathan, J.; Burrows, A.M.; Waller, B.M.; McComb, K. EquiFACS: The equine facial action coding system. *PLoS ONE* **2015**, *10*, e0131738. [CrossRef]
10. McLennan, K.M.; Rebelo, C.J.B.; Corke, M.J.; Holmes, M.A.; Leach, M.C.; Constantino-Casas, F. Development of a facial expression scale using footrot and mastitis as models of pain in sheep. *Appl. Anim. Behav. Sci.* **2016**, *176*, 19–26. [CrossRef]
11. Häger, C.; Biernot, S.; Buettner, M.; Glage, S.; Keubler, L.M.; Held, N.; Bleich, E.M.; Otto, K.; Müller, C.W.; Decker, S.; et al. The sheep grimace scale as an indicator of post-operative distress and pain in laboratory sheep. *PLoS ONE* **2017**, *12*, e0175839. [CrossRef] [PubMed]
12. Proctor, H.S.; Carder, G. Can ear postures reliably measure the positive emotional state of cows? *Appl. Anim. Behav. Sci.* **2014**, *161*, 20–27. [CrossRef]
13. Di Giminiani, P.; Brierley, V.L.M.H.; Scollo, A.; Gottardo, F.; Malcolm, E.M.; Edwards, S.A.; Leach, M.C. The assessment of facial expressions in piglets undergoing tail docking and castration: Toward the development of the piglet grimace scale. *Front. Vet. Sci.* **2016**, *3*, 100. [CrossRef] [PubMed]
14. Viscardi, A.V.; Hunniford, M.; Lawlis, P.; Leach, M.; Turner, P.V. Development of a piglet grimace scale to evaluate piglet pain using facial expressions following castration and tail docking: A pilot study. *Front. Vet. Sci.* **2017**, *4*, 51. [CrossRef]
15. Mota-Rojas, D.; Orihuela, A.; Martínez-Burnes, J.; Gómez, J.; Mora-Medina, P.; Alavez, B.; Ramírez, L.; González-Lozano, M. Neurological modulation of facial expressions in pigs and implications for production. *J. Anim. Behav. Biometeorol.* **2020**, *8*, 232–243. [CrossRef]
16. Descovich, K. Facial expression: An under-utilised tool for the assessment of welfare in mammals. *ALTEX* **2017**, *34*, 409–429. [CrossRef]
17. Waller, B.M.; Micheletta, J. Facial expression in nonhuman animals. *Emot. Rev.* **2013**, *5*, 54–59. [CrossRef]
18. Waller, B.M.; Peirce, K.; Caeiro, C.C.; Scheider, L.; Burrows, A.M.; McCune, S.; Kaminski, J. Paedomorphic facial expressions give dogs a selective advantage. *PLoS ONE* **2013**, *8*, e82686. [CrossRef] [PubMed]
19. Mariti, C.; Ricci, E.; Zilocchi, M.; Gazzano, A. Owners as a secure base for their dogs. *Behaviour* **2013**, *150*, 1275–1294. [CrossRef]
20. Hare, B.; Tomasello, M. Human-like social skills in dogs? *Trends Cogn. Sci.* **2005**, *9*, 439–444. [CrossRef]
21. Ford, G.; Guo, K.; Mills, D. Human facial expression affects a dog's response to conflicting directional gestural cues. *Behav. Process.* **2019**, *159*, 80–85. [CrossRef]
22. Katayama, M.; Kubo, T.; Yamakawa, T.; Fujiwara, K.; Nomoto, K.; Ikeda, K.; Mogi, K.; Nagasawa, M.; Kikusui, T. Emotional contagion from humans to dogs is facilitated by duration of ownership. *Front. Psychol.* **2019**, *10*, 1678. [CrossRef] [PubMed]
23. de Waal, F.B.M. What is an animal emotion? *Ann. N. Y. Acad. Sci.* **2011**, *1224*, 191–206. [CrossRef] [PubMed]
24. Ekman, P. Body position, facial expression, and verbal behavior during interviews. *J. Abnorm. Soc. Psych.* **1964**, *68*, 295–301. [CrossRef]
25. Bloom, T.; Friedman, H. Classifying dogs' (Canis Familiaris) facial expressions from photographs. *Behav. Proc.* **2013**, *96*, 1–10. [CrossRef]
26. Reid, J.; Nolan, A.M.; Scott, E.M. Measuring pain in dogs and cats using structured behavioural observation. *Vet. J.* **2018**, *236*, 72–79. [CrossRef] [PubMed]
27. Corke, M.J. Indicators of pain. In *Encyclopedia of Animal Behavior*, 2nd ed.; Choe, C.J., Ed.; Academic Press: Oxford, UK, 2019; Volume 1, pp. 147–152.

28. Lexis, H.; Weary, D.M. Facial Expressions in humans as a measure of empathy towards farm animals in pain. *PLoS ONE* **2021**, *16*, e0247808. [CrossRef]
29. McLennan, K.M.; Miller, A.L.; Dalla Costa, E.; Stucke, D.; Corke, M.J.; Broom, D.M.; Leach, M.C. Conceptual and methological issues relating to pain assessment in mammals: The development and ustilisation of pain facial expressions scales. *Appl. Anim. Behav. Sci.* **2019**, *217*, 1–15. [CrossRef]
30. Holden, E.; Calvo, G.; Collins, M.; Bell, A.; Reid, J.; Scott, E.M.; Nolan, A.M. Evaluation of facial expression in acute pain in cats. *J. Small Anim. Pract.* **2014**, *55*, 615–621. [CrossRef] [PubMed]
31. Evangelista, M.C.; Watanabe, R.; Leung, V.S.Y.; Monteiro, B.P.; O'Toole, E.; Pang, D.S.J.; Steagall, P.V. Facial Expressions of pain in cats: The development and validation of a feline grimace scale. *Sci. Rep.* **2019**, *9*, 19128. [CrossRef]
32. Langford, D.J.; Bailey, A.L.; Chanda, M.L.; Clarke, S.E.; Drummond, T.E.; Echols, S.; Glick, S.; Ingrao, J.; Klassen-Ross, T.; LaCroix-Fralish, M.L.; et al. Coding of facial expressions of pain in the laboratory mouse. *Nat. Methods* **2010**, *7*, 447–449. [CrossRef]
33. Sotocinal, S.G.; Sorge, R.E.; Zaloum, A.; Tuttle, A.H.; Martin, L.J.; Wieskopf, J.S.; Mapplebeck, J.C.; Wei, P.; Zhan, S.; Zhang, S.; et al. The rat grimace scale: A partially automated method for quantifying pain in the laboratory rat via facial expressions. *Mol. Pain* **2011**, *7*, 1744–8069. [CrossRef]
34. Lezama-García, K.; Orihuela, A.; Olmos-Hernández, A.; Reyes-Long, S.; Mota-Rojas, D. Facial expressions and emotions in domestic animals. *CAB. Rev.* **2019**, *14*, 1–12. [CrossRef]
35. Ekman, P. Are there basic emotions? *Psychol. Rev.* **1992**, *99*, 550–553. [CrossRef] [PubMed]
36. Harris, C.; Alvarado, N. Facial expressions, smile types, and self-report during humour, tickle, and pain. *Cogn. Emot.* **2005**, *19*, 655–669. [CrossRef]
37. Camerlink, I.; Coulange, E.; Farish, M.; Baxter, E.M.; Turner, S.P. Facial expression as a potential measure of both intent and emotion. *Sci. Rep.* **2018**, *8*, 17602. [CrossRef]
38. Gibson, J.E.; Pick, D.A. *An Ecological Approach to Perceptual Learning and Development*; Oxford University Press: New York, NY, USA, 2000; pp. 1–248.
39. Gaynor, J.S.; Muir, W.W. *Handbook of Veterinary Pain Management*, 3rd ed.; Elsevier: Missouri, MO, USA, 2015; pp. 1–627.
40. Siniscalchi, M.; D'Ingeo, S.; Minunno, M.; Quaranta, A. Communication in dogs. *Animals* **2018**, *8*, 131. [CrossRef]
41. Handelman, B. *Canine Behavior: A Photo Illustrated Handbook*; Dogwise Publishing: Wenatchee, WA, USA, 2012.
42. Hecht, J.; Horowitz, A. Introduction to dog behaviour. In *The Ethology of Domestic Animals: An Introductory Text*, 3rd ed.; Jensen, P., Ed.; CAB International: Oxford, UK, 2017; pp. 228–238.
43. Kaminski, J.; Hynds, J.; Morris, P.; Waller, B.M. Human attention affects facial expressions in domestic dogs. *Sci. Rep.* **2017**, *7*, 12914. [CrossRef] [PubMed]
44. Kunz, M.; Faltermeier, N.; Lautenbacher, S. Impact of visual learning on facial expressions of physical distress: A study on voluntary and evoked expressions of pain in congenitally blind and sighted individuals. *Biol. Psychol.* **2012**, *89*, 467–476. [CrossRef] [PubMed]
45. Bremhorst, A.; Mills, D.S.; Stolzlechner, L.; Würbel, H.; Riemer, S. 'Puppy dog eyes' are associated with eye movements, not communication. *Front. Psychol.* **2021**, *12*, 568935. [CrossRef] [PubMed]
46. Ferretti, V.; Papaleo, F. Understanding others: Emotion recognition in humans and other animals. *Genes Brain Behav.* **2019**, *18*, e12544. [CrossRef]
47. Bremhorst, A.; Sutter, N.A.; Würbel, H.; Mills, D.S.; Riemer, S. Differences in facial expressions during positive anticipation and frustration in dogs awaiting a reward. *Sci. Rep.* **2019**, *9*, 19312. [CrossRef] [PubMed]
48. Panksepp, J. The basic emotional circuits of mammalian brains: Do animals have affective lives? *Neurosci. Biobehav. Rev.* **2011**, *35*, 1791–1804. [CrossRef] [PubMed]
49. Cano-Vindel, A. Orientaciones en el estudio de la emoción. In *Manual de Motivación y Emoción*, 1st ed.; Fernández, E., Ed.; Editorial Universitaria Ramón Areces: Madrid, Spain, 1995; pp. 337–383.
50. Domínguez-Oliva, A.; Mota-Rojas, D.; Ruiz-García, A.G.; Miranda-Cortés, Á.E.; Hernández-Avalos, I. Clinical recognition of stress in dog and cats. *AMMVEPE* **2021**, *32*, 24–35.
51. Dawson, L.C.; Cheal, J.; Niel, L.; Mason, G. Humans can identify cats' affective states from subtle facial expressions. *Anim. Welf.* **2019**, *28*, 519–531. [CrossRef]
52. Andersen, P.H.; Broomé, S.; Rashid, M.; Lundblad, J.; Ask, K.; Li, Z.; Hernlund, E.; Rhodin, M.; Kjellström, H. Towards machine recognition of facial expressions of pain in horses. *Animals* **2021**, *11*, 1643. [CrossRef]
53. Urrego, D.; Múnera, A.; Troncoso, J. Peripheral facial nerve lesion induced long-term dendritic retraction in pyramidal cortico-facial neurons. *Biomed. Rev. Inst. Nac. Salud* **2011**, *31*, 560–569. [CrossRef]
54. Gil, V.A. *Anatomía del Sistema Nervioso en el Perro y en el Gato*; Servei de Publicacions Universitat Autónoma de Barcelona: Barcelona, Spain, 2008; pp. 1–263.
55. Castillo, G.D.; de Jorge, J.L. *Anatomía y Fisiología del Sistema Nervioso Central*, 1st ed.; Fundación Universitaria San Pablo CEU: Madrid, Spain, 2015; pp. 9–699.
56. Holmberg, J. The secretory nerves of the parotid gland of the dog. *J. Physiol.* **1971**, *219*, 463–476. [CrossRef]
57. Waller, B.M.; Caeiro, C.; Peirce, K.; Burrows, A.M.; Kaminski, J. *DogFACS: The Dog Facial Action Coding System Manual*; University of Portsmouth: Cambridge, UK, 2013; pp. 1–62.

58. Popesko, P. *Atlas de Anatomía Topográfica en los Animales Domésticos*, 2nd ed.; Masson, S.A., Ed.; En Stock: Barcelona, Spain, 1998; Volumen II, pp. 1–190.
59. López, P.C.; Mayor, A.P.; Labeaga, J.R.; López, B.M.; Pereira, T.H.D.S.; Monteiro, F.O.B. *Atlas de Los Músculos del Perro*; Edufra: Terra Firme, Brazil, 2018; pp. 1–128.
60. Burrows, A.M.; Kaminski, J.; Waller, B.M.; Omstead, K.M.; Rogers-Vizena, C.; Mendelson, B. Dog faces exhibit anatomical differences in comparison to other domestic animals. *Anat. Rec.* **2021**, *304*, 231–241. [CrossRef]
61. Vick, S.-J.; Waller, B.M.; Parr, L.A.; Smith Pasqualini, M.C.; Bard, K.A. A Cross-species comparison of facial morphology and movement in humans and chimpanzees using the Facial Action Coding System (FACS). *J. Nonverbal. Behav.* **2007**, *31*, 1–20. [CrossRef]
62. Caeiro, C.C.; Burrows, A.; Waller, B.M. Development and application of CatFACS: Are human cat adopters influenced by cat facial expressions? *Appl. Anim. Behav. Sci.* **2017**, *189*, 66–78. [CrossRef]
63. Kaminski, J.; Waller, B.M.; Diogo, R.; Hartstone-Rose, A.; Burrows, A.M. Evolution of facial muscle anatomy in dogs. *Proc. Natl. Acad. Sci. USA* **2019**, *116*, 14677–14681. [CrossRef] [PubMed]
64. Sternglanz, S.H.; Gray, J.L.; Murakami, M. Adult preferences for infantile facial features: An ethological approach. *Anim. Behav.* **1977**, *25*, 108–115. [CrossRef]
65. Archer, J.; Monton, S. Preferences for infant facial features in pet dogs and cats. *Ethology* **2011**, *118*, 217–226. [CrossRef]
66. Little, A.C. Manipulation of infant-like traits affects perceived cuteness of infant, adult and cat faces. *Ethology* **2012**, *118*, 775–782. [CrossRef]
67. Correira-Caeiro, C.; Guo, K.; Mills, D.S. Perception of dynamic facial expressions of emotion between dogs and humans. *Anim. Cogn.* **2020**, *23*, 465–476. [CrossRef] [PubMed]
68. Erkoç, T.; Ağdoğan, D.; Eskil, M.T. An observation based muscle model for simulation of facial expressions. *Signal Process. Image Commun.* **2018**, *64*, 11–20. [CrossRef]
69. Fenton, B.W.; Shih, E.; Zolton, J. The neurobiology of pain perception in normal and persistent pain. *Pain Manag.* **2015**, *5*, 297–317. [CrossRef]
70. Mota-Rojas, D.; Mariti, C.; Zdeinert, A.; Riggio, G.; Mora-Medina, P.; del Mar Reyes, A.; Gazzano, A.; Domínguez-Oliva, A.; Lezama-García, K.; José-Pérez, N.; et al. Anthropomorphism and Its Adverse Effects on the Distress and Welfare of Companion Animals. *Animals* **2021**, *11*, 3263. [CrossRef]
71. Phillips, M.L. Understanding the neurobiology of emotion perception: Implications for psychiatry. *Br. J. Psychiatry* **2003**, *182*, 190–192. [CrossRef]
72. Calder, A.J.; Lawrence, A.D.; Young, A.W. Neuropsychology of fear and loathing. *Nat. Rev. Neurosci.* **2001**, *2*, 352–363. [CrossRef]
73. Pertovaara, A. The noradrenergic pain regulation system: A potential target for pain therapy. *Eur. J. Pharmacol.* **2013**, *716*, 2–7. [CrossRef] [PubMed]
74. Reid, K.; Rogers, C.W.; Gronqvist, G.; Gee, E.K.; Bolwell, C.F. Anxiety and pain in horses measured by heart rate variability and behavior. *J. Vet. Behav.* **2017**, *22*, 1–6. [CrossRef]
75. Hu, J.; Qi, S.; Becker, B.; Luo, L.; Gao, S.; Gong, Q.; Hurlemann, R.; Kendrick, K.M. Oxytocin selectively facilitates learning with social feedback and increases activity and functional connectivity in emotional memory and reward processing regions. *Hum. Brain Mapp.* **2015**, *36*, 2132–2146. [CrossRef] [PubMed]
76. Valenchon, M.; Lévy, F.; Moussu, C.; Lansade, L. Stress Affects instrumental learning based on positive or negative reinforcement in interaction with personality in domestic horses. *PLoS ONE* **2017**, *12*, e0170783. [CrossRef] [PubMed]
77. Powell, L.; Guastella, A.J.; McGreevy, P.; Bauman, A.; Edwards, K.M.; Stamatakis, E. The physiological function of oxytocin in humans and its acute response to human-dog interactions: A review of the literature. *J. Vet. Behav.* **2019**, *30*, 25–32. [CrossRef]
78. Alexander, R.; Aragón, O.R.; Bookwala, J.; Cherbuin, N.; Gatt, J.M.; Kahrilas, I.J.; Kästner, N.; Lawrence, A.; Lowe, L.; Morrison, R.G.; et al. The neuroscience of positive emotions and affect: Implications for cultivating happiness and wellbeing. *Neurosci. Biobehav. Rev.* **2021**, *121*, 220–249. [CrossRef]
79. Osella, M.; Odore, R.; Badino, P.; Cuniberti, B.; Bergamasco, L. Plasma dopamine neurophysiological correlates in anxious dogs. In *Current Issues and Research in Veterinary Behavioral Medicine*; Purdue University Press: West Lafayette, IN, USA, 2005; pp. 274–276.
80. Karpiński, M.; Ognik, K.; Garbiec, A.; Czyżowski, P.; Krauze, M. Effect of stroking on serotonin, noradrenaline, and cortisol levels in the blood of right- and left-pawed dogs. *Animals* **2021**, *11*, 331. [CrossRef] [PubMed]
81. Siracusa, C. IAggression–Dogs. In *Small Animal Veterinary Psychiatry*; Denenberg, S., Ed.; CAB International: Oxfordshire, UK, 2021; pp. 191–193.
82. Bethlehem, R.A.I.; van Honk, J.; Auyeung, B.; Baron-Cohen, S. Oxytocin, brain physiology, and functional connectivity: A review of intranasal oxytocin FMRI studies. *Psychoneuroendocrinology* **2013**, *38*, 962–974. [CrossRef] [PubMed]
83. Domes, G.; Heinrichs, M.; Gläscher, J.; Büchel, C.; Braus, D.F.; Herpertz, S.C. Oxytocin attenuates amygdala responses to emotional faces regardless of valence. *Biol. Psychiatry* **2007**, *62*, 1187–1190. [CrossRef]
84. Petersson, M.; Uvnäs-Moberg, K.; Nilsson, A.; Gustafson, L.L.; Hydbring-Sandberg, E.; Handlin, L. Oxytocin and cortisol levels in dog owners and their dogs are associated with behavioral patterns: An exploratory study. *Front. Psychol.* **2017**, *8*, 1976. [CrossRef]
85. Mitsui, S.; Yamamoto, M.; Nagasawa, M.; Mogi, K.; Kikusui, T.; Ohtani, N.; Ohta, M. Urinary oxytocin as a noninvasive biomarker of positive emotion in dogs. *Horm. Behav.* **2011**, *60*, 239–243. [CrossRef] [PubMed]

86. Lansade, L.; Nowak, R.; Lainé, A.-L.; Leterrier, C.; Bonneau, C.; Parias, C.; Bertin, A. facial expression and oxytocin as possible markers of positive emotions in horses. *Sci. Rep.* **2018**, *8*, 14680. [CrossRef]
87. Ogi, A.; Mariti, C.; Baragli, P.; Sergi, V.; Gazzano, A. Effects of stroking on salivary oxytocin and cortisol in guide dogs: Preliminary results. *Animals* **2020**, *10*, 708. [CrossRef] [PubMed]
88. MacLean, E.L.; Gesquiere, L.R.; Gee, N.R.; Levy, K.; Martin, W.L.; Carter, C.S. Effects of affiliative human-animal interaction on dog salivary and plasma oxytocin and vasopressin. *Front. Psychol.* **2017**, *8*, 1606. [CrossRef]
89. Somppi, S.; Törnqvist, H.; Topál, J.; Koskela, A.; Hänninen, L.; Krause, C.M.; Vainio, O. Nasal oxytocin treatment biases dogs' visual attention and emotional response toward positive human facial expressions. *Front. Psychol.* **2017**, *8*, 1854. [CrossRef]
90. Mobbs, D. What Can the social emotions of dogs teach us about human emotions? *Anim. Sentience* **2018**, *3*, 5. [CrossRef]
91. Anderson, D.J.; Adolphs, R. A Framework for studying emotions across species. *Cell* **2014**, *157*, 187–200. [CrossRef]
92. Lindsley, D.B. Emotion. In *Handbook of Experimental Psychology*; Stevens, S.S., Ed.; Wiley: Oxford, UK, 2018; pp. 473–516.
93. Nummenmaa, L.; Saarimäki, H. Emotions as discrete patterns of systemic activity. *Neurosci. Lett.* **2019**, *693*, 3–8. [CrossRef]
94. Hess, U.; Hareli, S. The role of social context for the interpretation of emotional facial expressions. In *Understanding Facial Expressions in Communication*; Springer: New Delhi, India, 2015; pp. 119–141.
95. Kraaijenvanger, E.J.; Hofman, D.; Bos, P.A. A neuroendocrine account of facial mimicry and its dynamic modulation. *Neurosci. Biobehav. Rev.* **2017**, *77*, 98–106. [CrossRef]
96. Brudzynski, S.M. Communication of emotions in animals. In *Encyclopedia of Behavioral Neuroscience*; Elsevier Science: Amsterdam, The Netherlands, 2010; pp. 302–307.
97. Nagasawa, M.; Murai, K.; Mogi, K.; Kikusui, T. Dogs can discriminate human smiling faces from blank expressions. *Anim. Cogn.* **2011**, *14*, 525–533. [CrossRef]
98. Albuquerque, N.; Guo, K.; Wilkinson, A.; Resende, B.; Mills, D.S. Mouth-licking by dogs as a response to emotional stimuli. *Behav. Proc.* **2018**, *146*, 42–45. [CrossRef]
99. Karl, S.; Sladky, R.; Lamm, C.; Huber, L. Neural responses of pet dogs witnessing their caregiver´s positive interactions with a conspecific: An fMRI study. *Cereb. Cortex Commun.* **2021**, *2*, tgab047. [CrossRef]
100. Kujala, M.; Somppi, S.; Jokela, M.; Vainio, O.; Parkkonen, L. Human empathy, personality and experience affect the emotion ratings of dog and human facial expressions. *PLoS ONE* **2017**, *12*, e0170730. [CrossRef]
101. Beerda, B.; Schilder, M.B.H.; van Hooff, J.A.R.A.M.; de Vries, H.W. Manifestations of chronic and acute stress in dogs. *Appl. Anim. Behav. Sci.* **1997**, *52*, 307–319. [CrossRef]
102. Caeiro, C.; Guo, K.; Mills, D. Dogs and humans respond to emotionally competent stimuli by producing different facial actions. *Sci. Rep.* **2017**, *7*, 15525. [CrossRef] [PubMed]
103. Meints, K.; Racca, A.; Hickey, N. How to prevent dog bite injuries? Children misinterpret dogs, facial expressions. In Proceedings of the 10th World Conference on Injury Prevention and Safety Promotion, London, UK, 21–24 September 2010; BMJ Press: London, UK, 2010.
104. Bloom, T.; Trevathan-Minnis, M.; Atlas, N.; MacDonald, D.A.; Friedman, H.L. Identifying facial expressions in dogs: A replication and extension study. *Behav. Proc.* **2021**, *186*, 104371. [CrossRef]
105. Boissy, A.; Dwyer, C.M.; Jones, R.B. Fear and other negative emotions. In *Animal Welfare*; CAB International: Wallingford, UK, 2011; pp. 92–113.
106. Flint, H.E.; Coe, J.B.; Pearl, D.L.; Serpell, J.A.; Niel, L. Effect of training for dog fear identification on dog owner ratings of fear in familiar and unfamiliar dogs. *Appl. Anim. Behav. Sci.* **2018**, *208*, 66–74. [CrossRef]
107. Mariti, C.; Gazzano, A.; Moore, J.L.; Baragli, P.; Chelli, L.; Sighieri, C. Perception of dogs´stress their owners. *J. Vet. Behav.* **2012**, *7*, 213–219. [CrossRef]
108. Racca, A.; Amadei, E.; Ligout, S.; Guo, K.; Meints, K.; Mills, D. Discrimination of human and dog faces and inversion responses in domestic dogs (Canis familiaris). *Anim. Cogn.* **2010**, *13*, 525–533. [CrossRef] [PubMed]
109. Racca, A.; Guo, K.; Meints, K.; Mills, D.S. Reading faces: Differential Lateral gaze bias in processing canine and human facial expressions in dogs and 4-year-old children. *PLoS ONE* **2012**, *7*, e36076. [CrossRef]
110. Stetina, B.U.; Turner, K.; Burger, E.; Glenk, L.M.; McElheney, J.C.; Handlos, U.; Kothgassner, O.D. Learning emotion recognition from canines? Two for the road. *J. Vet. Behav.* **2011**, *6*, 108–114. [CrossRef]
111. Ogura, T.; Maki, M.; Nagata, S.; Nakamura, S. Dogs (Canis Familiaris) gaze at our hands: A preliminary eye-tracker experiment on selective attention in dogs. *Animals* **2020**, *10*, 755. [CrossRef] [PubMed]
112. Travain, T.; Valsecchi, P. Infrared thermography in the study of animals´emotional responses: A critical review. *Animals* **2021**, *11*, 2510. [CrossRef] [PubMed]
113. Langford, D.J. Social modulation of pain as evidence for empathy in mice. *Science* **2006**, *312*, 1967–1970. [CrossRef]
114. Jeon, D.; Kim, S.; Chetana, M.; Jo, D.; Ruley, H.E.; Lin, S.-Y.; Rabah, D.; Kinet, J.-P.; Shin, H.-S. Observational fear learning involves affective pain system and Cav1.2 Ca^{2+} channels in ACC. *Nat. Neurosci.* **2010**, *13*, 482–488. [CrossRef]
115. Hess, U.; Fischer, A. Emotional mimicry as social regulation. *Personal. Soc. Psychol. Rev.* **2013**, *17*, 142–157. [CrossRef] [PubMed]
116. Ha, J.C.; Campion, T.L. The emotional animal: Using the science of emotions to interpret behavior. In *Dog Behavior*; Academic Press: London, UK, 2019; pp. 93–108.
117. Kurachi, T.; Irimajiri, M.; Mizuta, Y.; Satoh, T. Dogs predisposed to anxiety disorders and related factors in japan. *Appl. Anim. Behav. Sci.* **2017**, *196*, 69–75. [CrossRef]

118. Van Bourg, J.; Patterson, J.E.; Wynne, C.D.L. Pet dogs (Canis Lupus Familiaris) release their trapped and distressed owners: Individual variation and evidence of emotional contagion. *PLoS ONE* **2020**, *15*, e0231742. [CrossRef] [PubMed]
119. Marcet-Rius, M.; Pageat, P.; Bienboire-Frosini, C.; Teruel, E.; Monneret, P.; Leclercq, J.; Lafont-Lecuelle, C.; Cozzi, A. Tail and ear movements as possible indicators of emotions in pigs. *Appl. Anim. Behav. Sci.* **2018**, *205*, 14–18. [CrossRef]
120. Reefmann, N.; Bütikofer Kaszàs, F.; Wechsler, B.; Gygax, L. Ear and tail postures as indicators of emotional valence in sheep. *Appl. Anim. Behav. Sci.* **2009**, *118*, 199–207. [CrossRef]
121. Briefer Freymond, S.; Briefer, E.F.; Zollinger, A.; Gindrat-von Allmen, Y.; Wyss, C.; Bachmann, I. Behaviour of horses in a judgment bias test associated with positive or negative reinforcement. *Appl. Anim. Behav. Sci.* **2014**, *158*, 34–45. [CrossRef]
122. Arena, L.; Wemelsfelder, F.; Messori, S.; Ferri, N.; Barnard, S. Application of free choice profiling to assess the emotional state of dogs housed in shelter environments. *Appl. Anim. Behav. Sci.* **2017**, *195*, 72–79. [CrossRef]
123. Tami, G.; Gallagher, A. Description of the behaviour of domestic dog (Canis Familiaris) by experienced and inexperienced people. *Appl. Anim. Behav. Sci.* **2009**, *120*, 159–169. [CrossRef]
124. Diesel, G.; Brodbelt, D.; Pfeiffer, U.D. Reliability of assessment of dogs´behavioural responses by staff working at a welfare charuty in the UK. *Appl. Anim. Behav. Sci.* **2008**, *115*, 171–181. [CrossRef]
125. Lakestani, N.N.; Donaldson, M.L.; Waran, N. Interpretation of dog behavior by children and young adults. *Anthrozoös* **2014**, *27*, 65–80. [CrossRef]
126. Merola, I.; Prato-Previde, E.; Marshall-Pescini, S. Dogs' social referencing towards owners and strangers. *PLoS ONE* **2012**, *7*, e47653. [CrossRef]
127. Horowitz, A. Disambiguating the "Guilty Look": Salient prompts to a familiar dog behaviour. *Behav. Proc.* **2009**, *81*, 447–452. [CrossRef]
128. Kujala, M. Canine Emotions as seen through human social cognition. *Anim. Sentience* **2017**, *2*, 1. [CrossRef]
129. Correia-Caeiro, C.; Guo, K.; Mills, D. Bodily emotional expressions are a primary source of information for dogs, but not for humans. *Anim. Cogn.* **2021**, *24*, 267–279. [CrossRef]
130. Csoltova, E.; Mehinagic, E. Where do we stand in the domestic dog (Canis familiaris) positive-emotion assessment: A state-of-the-art review and future directions. *Front. Psychol.* **2020**, *11*, 2131. [CrossRef]
131. Bremhorst, A.; Bütler, S.; Würbel, H.; Riemer, S. Incentive motivation in pet dogs–Preference for constant vs varied food rewards. *Sci. Rep.* **2018**, *8*, 9756. [CrossRef] [PubMed]
132. Siniscalchi, M.; d'Ingeo, S.; Quaranta, A. Orienting asymmetries and physiological reactivity in dogs' response to human emotional faces. *Learn. Behav.* **2018**, *46*, 574–585. [CrossRef] [PubMed]
133. Firnkes, A.; Bartels, A.; Bidoli, E.; Erhard, M. Appeasement signals used by dogs during dog–human communication. *J. Vet. Behav.* **2017**, *19*, 35–44. [CrossRef]
134. Nagasawa, M.; Kawai, E.; Mogi, K.; Kikusui, T. Dogs show left facial lateralization upon reunion with their owners. *Behav. Proc.* **2013**, *98*, 112–116. [CrossRef] [PubMed]
135. Siniscalchi, M.; D'Ingeo, S.; Fornelli, S.; Quaranta, A. Are dogs red–green colour blind? *R. Soc. Open Sci.* **2017**, *4*, 170869. [CrossRef]
136. Raja, S.N.; Carr, D.B.; Cohen, M.; Finnerup, N.B.; Flor, H.; Gibson, S.; Keefe, F.J.; Mogil, J.S.; Ringkamp, M.; Sluka, K.A.; et al. The revised international association for the study of pain definition of pain: Concepts, challenges, and compromises. *Pain* **2020**, *161*, 1976–1982. [CrossRef] [PubMed]
137. Mota-Rojas, D.; Orihuela, A.; Strappini-Asteggiano, A.; Nelly Cajiao-Pachón, M.; Agüera-Buendía, E.; Mora-Medina, P.; Ghezzi, M.; Alonso-Spilsbury, M. Teaching Animal Welfare in Veterinary Schools in Latin America. *Int. J. Vet. Sci. Med.* **2018**, *6*, 131–140. [CrossRef] [PubMed]
138. Hernandez-Avalos, I.; Mota-Rojas, D.; Mora-Medina, P.; Martínez-Burnes, J.; Casas-Alvarado, A.; Verduzco-Mendoza, A.; Lezama-García, K.; Olmos-Hernandez, A. Review of different methods used for clinical recognition and assessment of pain in dogs and cats. *Int. J. Vet. Sci. Med.* **2019**, *7*, 43–54. [CrossRef]
139. de Grauw, J.C.; van Loon, J.P.A.M. Systematic pain assessment in horses. *Vet. J.* **2016**, *209*, 14–22. [CrossRef]
140. Gleerup, K.B.; Forkman, B.; Lindegaard, C.; Andersen, P.H. An equine pain face. *Vet. Anaesth. Anal.* **2015**, *42*, 103–114. [CrossRef]
141. Hernández-Avalos, I.; Mota Rojas, D.; Mendoza-Flores, J.E.; Casas-Alvarado, A.; Flores-Padilla, K.; Miranda-Cortes, A.E.; Torres-Bernal, F.; Gómez-Prado, J.; Mora-Medina, P. Nociceptive pain and anxiety in equines: Physiological and behavioral alterations. *Vet. World* **2021**, in press.
142. Dalla Costa, E.; Stucke, D.; Dai, F.; Minero, M.; Leach, M.; Lebelt, D. Using the horse grimace scale (HGS) to assess pain associated with acute laminitis in horses (*Equus caballus*). *Animals* **2016**, *6*, 47. [CrossRef]
143. Parr, L.A.; Waller, B.M.; Burrows, A.M.; Gothard, K.M.; Vick, S.J. Brief communication: MaqFACS: A muscle-based facial movement coding system for the rhesus macaque. *Am. J. Phys. Anthropol.* **2010**, *143*, 625–630. [CrossRef] [PubMed]
144. Keating, S.C.J.; Thomas, A.A.; Flecknell, P.A.; Leach, M.C. Evaluation of EMLA cream for preventing pain during tattooing of rabbits: Changes in physiological, behavioural and facial expression responses. *PLoS ONE* **2012**, *7*, e44437. [CrossRef]
145. Reijgwart, M.L.; Schoemaker, N.J.; Pascuzzo, R.; Leach, M.C.; Stodel, M.; de Nies, L.; Hendriksen, C.F.M.; van der Meer, M.; Vinke, C.M.; van Zeeland, Y.R.A. The composition and initial evaluation of a grimace scale in ferrets after surgical implantation of a telemetry probe. *PLoS ONE* **2017**, *12*, e0187986. [CrossRef] [PubMed]

146. Camps, T.; Amat, M.; Mariotti, V.M.; Le Brech, S.; Manteca, X. Pain-related aggression in dogs: 12 clinical cases. *J. Vet. Behav.* **2012**, *7*, 99–102. [CrossRef]
147. Dalla Costa, E.; Minero, M.; Lebelt, D.; Stucke, D.; Canali, E.; Leach, M.C. Development of the horse grimace scale (HGS) as a pain assessment tool in horses undergoing routine castration. *PLoS ONE* **2014**, *9*, e92281. [CrossRef]
148. Niella, R.V.; Sena, A.S.; Corrêa, J.M.X.; Soares, P.C.L.R.; Pinto, T.M.; Silva Junior, A.C.; Costa, B.A.; de Oliveira, J.N.S.; da Silva, E.B.; de Lavor, M.S.L. Preemptive effect of amantadine as adjuvant in postoperative analgesia of ovaryhisterectomy in dogs. *Res. Soc. Dev.* **2020**, *9*, e68091110128. [CrossRef]
149. Barletta, M.; Young, C.N.; Quandt, J.E.; Hofmeister, E.H. Agreement between veterinary students and anesthesiologists regarding postoperative pain assessment in dogs. *Vet. Anaesth. Analg.* **2016**, *43*, 91–98. [CrossRef]
150. Kinjavdekar, S.P.; Aithal, A.H.P.; Pawde, A.M.; Malik, V. Comparison of analgesic effects of meloxicam and ketoprofen using university of melbourne pain scale in clinical canine orthopaedic patients. *J. Appl. Anim. Res.* **2010**, *38*, 261–264. [CrossRef]
151. Reader, R.C.; McCarthy, R.J.; Schultz, K.L.; Volturo, A.R.; Barton, B.A.; O'Hara, M.J.; Abelson, A.L. Comparison of liposomal bupivacaine and 0.5% bupivacaine hydrochloride for control of postoperative pain in dogs undergoing tibial plateau leveling osteotomy. *J. Am. Vet. Med. Assoc.* **2020**, *256*, 1011–1019. [CrossRef]
152. Brondani, J.T.; Mama, K.R.; Luna, S.P.L.; Wright, B.D.; Niyom, S.; Ambrosio, J.; Vogel, P.R.; Padovani, C.R. Validation of the English Version of the UNESP-Botucatu Multidimensional Composite Pain Scale for Assessing Postoperative Pain in Cats. *BMC Vet. Res.* **2013**, *9*, 143. [CrossRef]
153. Belli, M.; de Oliveira, A.R.; de Lima, M.T.; Trindade, P.H.E.; Steagall, P.V.; Luna, S.P.L. Clinical validation of the Short and Long UNESP-Botucatu Scales for Feline Pain Assessment. *PeerJ* **2021**, *9*, e11225. [CrossRef] [PubMed]
154. Meunier, N.V.; Panti, A.; Mazeri, S.; Fernandes, K.A.; Handel, I.G.; Bronsvoort, B.M.d.C.; Gamble, L.; Mellanby, R.J. Randomised trial of perioperative tramadol for canine sterilisation pain management. *Vet. Rec.* **2019**, *185*, 406. [CrossRef] [PubMed]
155. Reid, J.; Scott, E.M.; Calvo, G.; Nolan, A.M. Definitive Glasgow acute pain scale for cats: Validation and intervention level. *Vet. Rec.* **2017**, *180*, 449. [CrossRef]
156. Steagall, P.V.; Beatriz, M.P. Acute pain in cats recent advances in clinical assessment. *J. Feline Med. Surg.* **2019**, *21*, 25–34. [CrossRef]
157. Mich, P.M.; Hellyer, P.W. Objective, categoric methods for assessing pain and analgesia. In *Handbook of Veterinary Pain Management*, 2nd ed.; Gaynor, J.S., Muir, W.W., Eds.; Mosby: St. Louis, MO, USA, 2008; pp. 78–109.
158. Lu, Y.; Mahmoud, M.; Robinson, P. Estimating sheep pain level using facial action unit detection. In Proceedings of the 12th IEEE International Conference on Automatic Face and Gesture Recognition, Cambridge, UK, 30 May–3 June 2017; pp. 394–399. [CrossRef]
159. Neethirajan, S.; Reimert, I.; Kemp, B. Measuring farm animal emotions—Sensor-based approaches. *Sensors* **2021**, *21*, 553. [CrossRef]
160. Casas-Alvarado, A.; Mota-Rojas, D.; Hernández-Avalos, I.; Mora-Medina, P.; Olmos-Hernández, A.; Verduzco-Mendoza, A.; Reyes-Sotelo, B.; Martínez-Burnes, J. Advances in infrared thermography: Surgical aspects, vascular changes, and pain monitoring in veterinary medicine. *J. Therm. Biol.* **2020**, *92*, 102664. [CrossRef] [PubMed]
161. Mota-Rojas, D.; Pereira, A.; Wang, D.; Martínez-Burnes, J.; Ghezzi, M.; Lendez, P.; Bertoni, A.; Geraldo, A.M. Clinical applications and factors involved in validating thermal windows used in infrared thermography to assess health and productivity. *Animals* **2021**, *11*, 2247. [CrossRef]
162. Reyes-Sotelo, B.; Mota-Rojas, D.; Martínez-Burnes, J.; Gómez, J.; Lezama, K.; González-Lozano, M.; Hernández-Ávalos, I.; Casas, A.; Herrera, Y.; Mora-Medina, P. Tail docking in dogs: Behavioural, physiological and ethical aspects. *CAB Rev.* **2020**, *15*, 1–13.
163. Mota-Rojas, D.; Broom, D.M.; Orihuela, A.; Velarde, A.; Napolitano, F.; Alonso-Spilsbury, M. Effects of human-animal relationship on animal welfare. *J. Anim. Behav. Biometeorol.* **2020**, *8*, 196–205. [CrossRef]

MDPI

Review

Thermal and Circulatory Changes in Diverse Body Regions in Dogs and Cats Evaluated by Infrared Thermography

Alejandro Casas-Alvarado [1], Julio Martínez-Burnes [2], Patricia Mora-Medina [3], Ismael Hernández-Avalos [3], Adriana Domínguez-Oliva [4], Karina Lezama-García [4], Jocelyn Gómez-Prado [4] and Daniel Mota-Rojas [4,*]

1 Ph.D. Program in Biological and Health Sciences, [Doctorado en Ciencias Biológicas y de la Salud], Universidad Autónoma Metropolitana, Mexico City 04960, Mexico; ale0164g@hotmail.com
2 Animal Health Group, Facultad de Medicina Veterinaria y Zootecnia, Universidad Autónoma de Tamaulipas, Victoria City 87000, Mexico; jmburnes@docentes.uat.edu.mx
3 Facultad de Estudios Superiores Cuautitlán, Universidad Nacional Autónoma de Mexico (UNAM), Mexico City 54714, Mexico; mormed2001@yahoo.com.mx (P.M.-M.); mvziha@hotmail.com (I.H.-A.)
4 Neurophysiology, Behavior and Animal Welfare Assessment, DPAA, Universidad Autónoma Metropolitana (UAM), Xochimilco Campus, Mexico City 04960, Mexico; mvz.freena@gmail.com (A.D.-O.); kikislezama@hotmail.com (K.L.-G.); jocelyn.gomez.ilp@gmail.com (J.G.-P.)
* Correspondence: dmota100@yahoo.com.mx

Simple Summary: Infrared thermography is a tool that measures changes in the surface temperature of the skin and has been used in companion animals to determine their health state and to diagnose inflammatory processes, neoplasia, pain, or neuropathies. Body regions such as the face, body, or hind/forelimbs are commonly used for thermography in dogs and cats. However, since there is disagreement about the differences in temperature recorded using this tool, this article analyzes the usefulness of IRT in companion animals (pets) as a complementary diagnostic method for evaluating thermal and circulatory changes. It analyzes the recent scientific evidence on the use of facial, body, and appendicular thermal windows in dogs and cats under different clinical conditions.

Abstract: Infrared thermography (IRT) has been proposed as a method for clinical research to detect local inflammatory processes, wounds, neoplasms, pain, and neuropathies. However, evidence of the effectiveness of the thermal windows used in dogs and cats is discrepant. This review aims to analyze and discuss the usefulness of IRT in diverse body regions in household animals (pets) related to recent scientific evidence on the use of the facial, body, and appendicular thermal windows. IRT is a diagnostic method that evaluates thermal and circulatory changes under different clinical conditions. For the face, structures such as the lacrimal caruncle, ocular area, and pinna are sensitive to assessments of stress degrees, but only the ocular window has been validated in felines. The usefulness of body and appendicular thermal windows has not been conclusively demonstrated because evidence indicates that biological and environmental factors may strongly influence thermal responses in those body regions. The above has led to proposals to evaluate specific muscles that receive high circulation, such as the *biceps femoris* and *gracilis*. The neck area, perivulvar, and perianal regions may also prove to be useful thermal windows, but their degree of statistical reliability must be established. In conclusion, IRT is a non-invasive technique that can be used to diagnose inflammatory and neoplastic conditions early. However, additional research is required to establish the sensitivity and specificity of these thermal windows and validate their clinical use in dogs and cats.

Keywords: dogs; cats; thermal window; infrared thermography

Citation: Casas-Alvarado, A.; Martínez-Burnes, J.; Mora-Medina, P.; Hernández-Avalos, I.; Domínguez-Oliva, A.; Lezama-García, K.; Gómez-Prado, J.; Mota-Rojas, D. Thermal and Circulatory Changes in Diverse Body Regions in Dogs and Cats Evaluated by Infrared Thermography. *Animals* **2022**, *12*, 789. https://doi.org/10.3390/ani12060789

Academic Editor: Michael Lindinger

Received: 13 February 2022
Accepted: 16 March 2022
Published: 20 March 2022

1. Introduction

With the advent of new technologies to complement existing diagnostic methods, infrared thermography (IRT) has acquired greater importance in medicine and industry as a non-invasive method that does not use radiation [1,2]. IRT permits evaluating surface microcirculation based on consequent variations in tissue temperatures that medical research has associated with changes in the autonomic nervous system (ANS) of inflammatory, infectious, neoplastic, pain-induced, or stress-related origin [2–4]. Discussion of the medical use of IRT has centered on various anatomical regions that could be used as thermal windows to determine changes in surface blood flow that reflect an organism's autonomic activity. The term thermal window refers to regions with surface blood vessels where changes can be detected that represent surface circulation [2]. Regions identified to date include the lacrimal caruncle, eye, ear, thorax flanks, femoral area, and face [5–7].

Studies of structures such as the lacrimal caruncle and eye have established that IRT strongly correlates with the activity of the ANS [8–10]. When the sympathetic element of the ANS is stimulated, an initial vasoconstriction response is produced due to the neurosecretion of adrenaline and noradrenaline, causing an alteration in the surface thermal response [2,7]. These findings demonstrate IRT's usefulness for the early detection of inflammatory conditions [11], though its sensitivity is still in doubt because of potential biological and environmental influences [12]. This has been evidenced in appendicular regions, where sensitivity and effectiveness can be affected, for example, by the length or type of an animal's hair or fur [13].

In the face of discrepant evidence on the effectiveness of the thermal windows currently used with dogs and cats, the aim of this review is to discuss the usefulness of IRT in household animals (pets) as a diagnostic method for evaluating thermal and circulatory changes under different clinical conditions, and then to analyze recent scientific evidence on the use of the facial, body, and appendicular thermal windows in dogs and cats.

2. Clinical Usefulness of IRT with Dogs and Cats

The IRT technique detects the radiation spectrum from an object between 7.5 and 13 μm using a specialized camera that produces a thermal image visualized and interpreted using special software [14]. IRT makes it possible to visualize surface changes in skin temperature that can be correlated with the body region evaluated [1], for example, in cases of osteoarthritis [15], spinal injury [16], conduct pass evaluations [11,17], infectious diseases [18,19], and malignant neoplasias [20,21].

Contrasting from existing clinical methodologies, IRT has the advantage of its non-invasiveness to assess the thermal state of animals without the need for physical or chemical restraint [2]. It is considered a non-contact complementary technique that does not require setting devices on the animals' bodies [22] and may represent a useful method to avoid stress-induced hyperthermia due to handling [23,24]. Additionally, it is safe for animals [25], and the surface temperature changes can be evaluated in real-time with automated or handheld devices [26,27].

Hildebrandt et al. [28] observed that some types of lesions in local tissues are related to variations in blood flow that affect surface temperature, just as inflammation fosters hyperthermic states. IRT's capacity to detect temperature variations is useful for evaluating skin wounds or lesions and detecting changes associated with hypothermia—degeneration, reduced muscular activity, decreased blood perfusion—among other areas of study where findings suggest that it can be used as a diagnostic instrument. This is the case, as well, of some neurological disorders [29,30], surgeries [2], degrees of pain [31], urological problems [32], cancer [20], and infectious diseases [19].

IRT is not limited to identifying areas where local temperature increases, as in most inflammatory processes such as dermal lesions and wounds. It can also recognize changes associated with decreased temperature due to tissue degeneration or reduced muscle activity. In patients with neurological disorders, the decline in surface temperature in the site of injury [29,30], as well as the detection of hypothermia associated with a blood perfusion failure in individuals with thromboembolic cases, can be determined by IRT [33]. Similarly, the drop in temperature during tissue surgery can aid in identifying the viable conditions of a graft, hypoperfusion states [2], or estimate the presence and degree of pain [31,34].

Thermal windows are used to evaluate these states, where changes in blood perfusion allow the thermal exchange with the environment [3,35–37]. However, in companion animals, the most appropriate body regions to quantify the superficial thermal response have not yet been established [38] because the sensitivity and specificity of these regions can be affected by internal and external factors that alter the vascular modulation of the ANS towards different stimuli [14]. Due to discrepancies in the available literature regarding thermographic research on different anatomical areas, it is necessary to establish the sensitivity and specificity of these regions to determine whether they represent a tool to recognize surface microcirculation mediated by the ANS activity [14]. Therefore, the current literature regarding the use and possible validation of each region applied in companion animals will be discussed below.

3. Facial Thermal Windows

Studies of the facial region have described several thermal windows: the lacrimal caruncle [9,12,39,40], the eyeball [41], and the ear [22,42]. Some authors even suggest evaluating the entire area of the face [43]. In the cranial region (*regio cranii*), the auricula (*regio auricularis*) has been described as an area whose vascular response is associated with inflammatory, painful, and stressful events [22,42]. This region and the ocular window (*oculi*) in veterinary medicine are currently used as thermal windows to evaluate the infrared response [31,35,38].

The thermal window most widely analyzed in dogs is the caruncle (Figure 1) [44]. This region receives blood circulation from two main arteries, the supraorbital artery (*supraorbitalis*) and the infraorbital artery (*infraorbitalis*), from which emerges a capillary called the lacrimal (*lacrimalis*) that provides circulation to the lacrimal gland [45]. Its innervation is provided by the infraorbital nerve (*infraorbitalis*), derived as a branch of the facial nerve (*facialis*), considered a sensitive area that responds to the activity of the Sympathetic Nervous System (SNS) [41,46]. For this reason, the lacrimal caruncle has been extensively studied in animals to assess the ANS activity related to painful and stressful events [8,10,40,47,48].

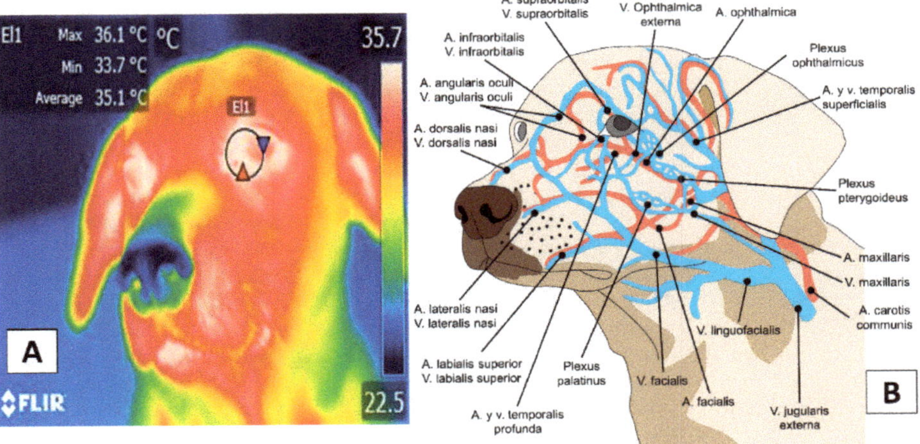

Figure 1. The thermal window of the lacrimal caruncle (El1). This window considers the region of the medial canthus of the upper and lower eyelids (**A**), which receive circulation from the supraorbital and infraorbital arteries (**B**). The latter holds sympathetic innervation from the infraorbital branch of the facial nerve, responsible for the autonomic hemodynamic activity of this region. Thermal images obtained using a FLIR thermal camera (Wilsonville, Oregon, U.S.).

On the other hand, the pinna or ear region presents a vasculature coming from the caudal auricular artery (*auricularis caudalis*), dividing into three branches: the lateral (*lateralis*), intermediate (*intermedius*), and medial (*medialis*). The branch of the facial nerve (*facialis*), known as auriculopalpebral (*auriculopalpebralis*), innervates this region [45], as shown in Figure 2 [49].

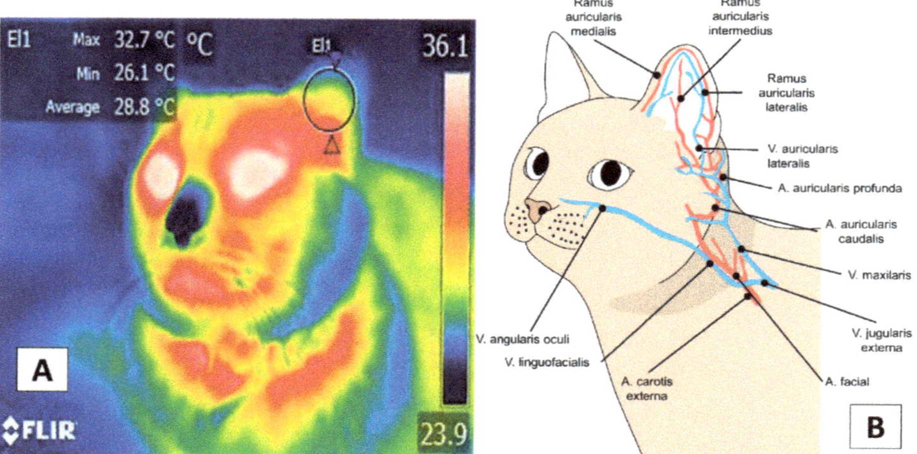

Figure 2. The thermal window of the ear. This window considers the auricular region marking a circle of approximately 2 cm (El1) to obtain the irradiated temperature of the tympanic membrane or the inner ear (**A**). This region presents irrigation from the deep auricular artery and three branches that allow thermal exchange with the environment (**B**) through vasomotor changes. Thermal images obtained using a FLIR thermal camera.

Travain et al. [8], for example, evaluated the usefulness of IRT for measuring degrees of stress in 14 dogs, reporting that the surface temperature of the lacrimal caruncle increased during the examination period, in contrast to the measurements obtained before and after the clinical check-up. They determined that a reduction in the level of activity had occurred, so this response was more sensitive for determining stress of psychogenic type. The biological explanation is that, during stress events, sympathetic activation causes an increase in temperature, known as stress hyperthermia [24], a phenomenon that is present in other species such as laboratory rats [50,51] and sheep [46]. In canines with separation anxiety, IRT applied in the same thermal window has detected an increase in ocular temperature when the owner is absent; however, an increase was also registered when the animal was in contact with the owner. These results show that, although ocular temperature responds to autonomic activity, it can have positive and negative valence [52].

This situation was discussed in a complementary study by Travain et al. [6], which analyzed the emotional responses of 19 dogs to potentially pleasant events (the presence of food in the form of candies or kibble) by evaluating their physiological (cardiac frequency) and cardiac responses (heart rate variability), and behavior, in addition to using IRT. Those authors observed that the temperature of the lacrimal caruncle and the cardiac frequency increased during positive stimulation but with no modification of heart rate variability. They also reported behaviors associated with excitation that occurred in the presence of the candies, manifested by greater tail movement.

The scientific evidence shows that the lacrimal caruncle's temperature has significantly increased due to the activity of the ANS associated with potentially stressful events, being a non-invasive indicator of this state or negative emotions such as anxiety or fear in companion animals [41,53]. Likewise, Travain et al. [54] mention that IRT on the ocular surface could also help study the temperament of animals due to the lateralization of blood circulation that results in different thermal patterns, depending on the reaction of the individual, the nature of the stressor, and the previous animal experiences.

There are no studies of canines similar to Stewart et al.'s work [47] with 30, 4-month-old bovines after castration, which established the presence of a cardiovascular response associated with pain due to activation of the SNS and found that the increase in the surface temperature of the lacrimal caruncle was related to higher blood pressure and cardiac frequency. Those authors also observed that this occurred synchronically with an increase in blood catecholamine levels, indicating that the thermal response of the lacrimal caruncle recorded by IRT is closely related to the activity of the ANS. The above mentioned is similar to observations of the authors in dogs (Figure 3). These results suggest that this technique and region could indirectly register autonomic activity in animals, unlike observations in humans [9].

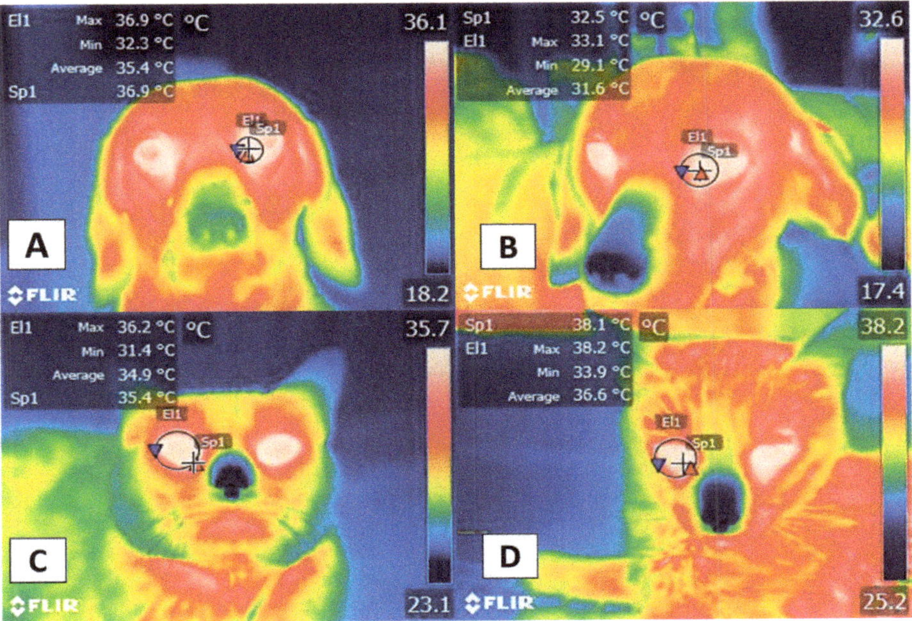

Figure 3. Facial thermal windows in the dog and cat: (**A**) Lacrimal caruncle. The lacrimal caruncle (El1) is shown in the eye region of a 4-year-old dog prior to surgery. This area runs from the medial commissure of the eyelid to the mid-region of the eye. Additionally shown is the temperature of a dog at rest. The maximum temperature was 36.9 °C (red triangle), the minimum was 32.3 °C (blue triangle). (**B**) Lacrimal caruncle (El1). This image shows modifications in the same dog after 2 h of concluding the surgical procedure with a decrease of 3.8 °C from the maximum (33.1 °C, red triangle) and 3.2 °C from the minimum (29.1 °C, blue triangle) generated by the perception of pain during the surgical procedure that activated the ANS, causing a reaction that consisted of peripheral vasoconstriction with a decrease in surface circulation that reduced the irradiated temperature. This would explain the thermal reaction observed. (**C**) Ocular region. Image showing the ocular thermal window (El1) of a 4-year-old feline wounded by the attack of a conspecific. This thermal window covers the entire interior zone of the eyeball. The maximum ocular temperature recorded was 37.3 °C (red triangle), the minimum was 33.9 °C (blue triangle) while the animal was comfortable or at rest. (**D**) There is an increase of 0.9 °C in the maximum temperature (38.2 °C, red triangle) due to peripheral vasodilation. This vasomotor response is a consequence of proinflammatory mediators (e.g., histamine) released after the injury. Sp1: default focal point of the software. Thermal images obtained using a FLIR thermal camera.

Zanghi [22] reported that the temperature of both the eye and auricular region showed a direct relation (r^2 = 0.67 eye, r = 0.61 ear) to rectal temperature in 32 dogs during exercise. They observed that rectal, eye, and auricular region temperatures decreased significantly as the exercise duration increased. These events differ to what Elias et al. [12] recently found when evaluating thermal stress and vascular responses before and after exercise in 465 greyhound dogs by recording average eye temperatures. They discovered that the eye temperature increased after exercise and, also, that the temperature of the right eye was more sensitive to changes but suggested the need to consider environmental and biological factors when detecting stress during exercise.

These findings prove that circulation in specific structures—especially the eye—responds more to ANS activity than the auricular region. Thus, it presents an effective technique for demonstrating degrees of stress in animals.

Regarding the relation of IRT to pain evaluation scales, the results are discrepant. For example, Lush and Ijichi [5] evaluated the behavioral and physiological responses of 20 dogs after orchiectomy. They affirmed that the temperature of the lacrimal caruncle could not be associated with pain scores (r = −0.107), though they did observe a temperature increase in that region. A separate study conducted on 30 bitches subjected to ovariohysterectomy under three epidural analgesia protocols reported that the temperature of the lacrimal caruncle did not show significant differences between groups. This effect was attributed to the analgesia provided, an element that was corroborated with the score obtained in two postoperative pain scales (the Dynamic Interactive Visual Analogue Scale and the University of Melbourne pain scale) [55]. While this led to the observation that the lacrimal caruncle presents vascular and thermal responses that accord with ANS activity, it raises questions about IRT's sensitivity and specificity when used in this region. It is important to note that some authors sustain that optimizing evaluations of thermal responses during painful events requires performing IRT in association with facial expressions. This approach helps determine the degree of pain during surgical processes or periods of intensive care. As expression intensity is accentuated, it can be correlated with thermal responses at a sensitivity of 75% [56,57].

Evidence for eye temperature relevance in cats is clear, at least according to Foster and Ijichi's [41] report on an evaluation of 34 cats in stress situations under distinct breeding conditions in which IRT was applied to the central region of the eye. Those authors observed that temperature correlated significantly with acceptable scores for the profile of feline temperament ($r^2 = -0.37$, $p = 0.028$) and established that eye temperatures were higher in older cats and those housed individually than those kept in groups.

Another field of evaluation of thermographic responses in cats involves stress induced by separation from the owner. Observations showed that the auricular region temperature decreased during separation but registered an increase as soon as the owner returned to contact with the patient [42]. The thermal window of the auricular pavilion (pinna) has also been used in companion animals to evaluate responses to stress in these species Reimer et al. [42] examined the ear temperature (in both ears) in six dogs subjected to a 2-minute-social isolation test from the owner. The temperature in both ears decreased by 0.2 to 0.6 °C during separation and increased in the owner's presence (between 0.4 and 0.7 °C). In contrast, the opposite effect has been observed in cats. In a study conducted on 41 domestic felines from one to eight years old, the animals were divided into two groups according to serum cortisol levels (high: 7 mcg/dL, and low: 3 mcg/dL) and were exposed to transport and an unknown environment to quantify the ear temperature. After the stimulus, the temperature (0.19 °C, $p < 0.04$), glucose, and cortisol increased significantly; however, a correlation between both factors could not be established. The authors also reported that the right ear but not the left ear temperature is related to stress-induced cortisolemia [58].

These results lead to questioning the usefulness of the thermal windows and IRT in general. As Polgár et al. [59] mentioned, thermography can be influenced by a wide variety of factors that could alter the accuracy to assess the valence of the emotional state of animals. IRT indicates a level of arousal and autonomic activity but does not assure that increases or decreases in temperature are attributed to a negative or positive stimulus. In contrast, in dogs, tympanic membrane temperature has been suggested as a suitable method to monitor temperature, as it is closely related to the standard method of rectal measurement, with a mean difference of 0.39 °C between both [60].

Sousa et al. [61] measured the ear temperature to estimate the standard rectal method in cats. In this study, 29 healthy cats were evaluated for two weeks, observing that ear temperature maintained a direct correlation with rectal temperature (r^2 = 0.96). Likewise, in 32 dogs, it has been observed that the ocular and auricular temperature maintained a direct correlation with the rectal temperature (r^2 = 0.67 eye, r^2 = 0.61 ear), considering the auricular region as an area that reflects accurately thermal rectal values [22]. In contrast, in a study conducted on 19 domestic cats, the ear and rectal temperatures had a weak correlation (r^2 = 0.62) with mean differences of 0.07 °C and limits of 1.43 °C [62]. These results are similar to those obtained by Nutt et al. [63], who evaluated the temperature in three anatomical regions (pinna, perineum, and gingiva) of 188 domestic shelter cats. A poor agreement was observed between ear IRT and rectal temperature using Bland–Altman analysis; furthermore, the mean values ranged from 0.7 to 1.3 °C. The researchers suggest that this tool does not offer reliable values to support its clinical use in the aforementioned regions. The main limitation reported in the ear is the presence of hair or secretions in the external auditory canal that can alter the values readings and interpretations [42,62,64].

The nasal plane has been used to determine the degree of comfort of canines undergoing radiotherapy for intranasal tumors. In a clinical case report of a shikoku inu dog (a breed native to Japan), nasal IRT showed an increase of up to 4.1 °C after radiotherapy, reaching a maximum temperature of 42.3 °C. In humans, the increase in skin temperature after such treatments is associated with stress and pain. However, in the shikoku inu dog used in the study by Saeki et al. [65], the lack of behaviors associated with pain or stress in the patient, and the late response of the increase in temperature (120–140 min after radiotherapy) did not establish a relationship between hyperthermia and pain. Despite these findings, other authors recommend evaluating the entire facial region, or the face itself, since observations in human medicine—according to recent reports based on assessments spanning the ears, nose, lips, and cheeks—could be associated with the degree of comfort that an individual feels at a precision as high as 85% [43,66]. Considering the entire facial region or specific thermal windows could represent a complimentary assessment through IRT. However, some environmental and technical factors must be considered before and during the application of IRT in animals.

Although the evidence and the use of IRT make it a useful complementary tool for clinical practice in veterinary medicine, in both humans and animals, environmental, individual, and technical aspects are reported as factors that can influence the values obtained by the IRT [67]. Solar radiation, humidity, and wind speed in animals can affect their reading. For example, direct solar radiation can increase the temperature of the different thermal windows by up to 0.5 °C, similar to what was reported by Loughmiller et al. [68], who observed in pigs that environmental temperature had a significant linear relationship with body surface temperature (slope value = 0.40 °C).

This environmental temperature influence has shown in equines a relationship with the IRT measured in the carpus region (r^2 = 0.88) [69], so that the climate conditions are a factor that, preferably, must be controlled to avoid variations. This has been applied in dogs by controlling the room temperature at 21 °C to unify the characteristics and obtain a reliable IRT reading [11].

The presence of wind is another factor that, contrary to solar radiation, can cause the ocular surface temperature to drop by 0.7 °C [70]. Therefore, these environmental factors can influence the IRT reading and must be considered or controlled by the evaluators for a correct results interpretation. In addition, biological elements such as the different species, the presence of hair, fur color, and the presence of fluids may also need to be evaluated, together with laterality.

The technical aspects are another variable that must be considered when using thermographic cameras. In humans, it is reported that technical factors such as the type of camera and the degree of resolution can affect the temperature measurement. According to some authors, the resolution of 320 × 240 pixels is recommended since it allows the detection of minimal changes in surface temperature, a process that is difficult with low image resolutions [2,15,71]. This factor is also closely related to the camera lens type since some lenses are coated with diamond or germanium films that allow infrared radiation to be captured, while other types of cameras with a compound lens system process the infrared radiation in an electronic circuit [2,72].

Therefore, it is important not only to establish the thermal window, recognizing that the eye region—the lacrimal caruncle or ocular area—is especially efficacious for determining the degree of thermal equilibrium under stress when compared to the ear, but also considering the type of camera and external factors that can alter IRT readings in companion animals.

4. Appendicular Windows

In addition to the facial thermal windows just discussed, the thoracic (*membri thoracici*) and pelvic limbs (*membri pelvini*) have been proposed for evaluating specific pathologies, such as inflammatory problems (Figure 4). The hip (*regio articulationis coxae*), shoulder (*regio articulationis humeri*), knees (*regio genus cranialis*), and elbows (*regio olecrani*) are regions susceptible to changes in the emitted radiation. In the case of the thoracic limb, this region is supplied by the axillary artery (*arteria axillaris*) emerging from the first rib and continuing as the deep brachial artery (*profunda brachii*); in the forearm as the deep ante-brachial artery (*profunda antebrachii*), and is divided into various branches at the distal level: in the palmar carpus (*ramus carpeus palmaris*) and the palmar (*ramus palmaris*) [45].

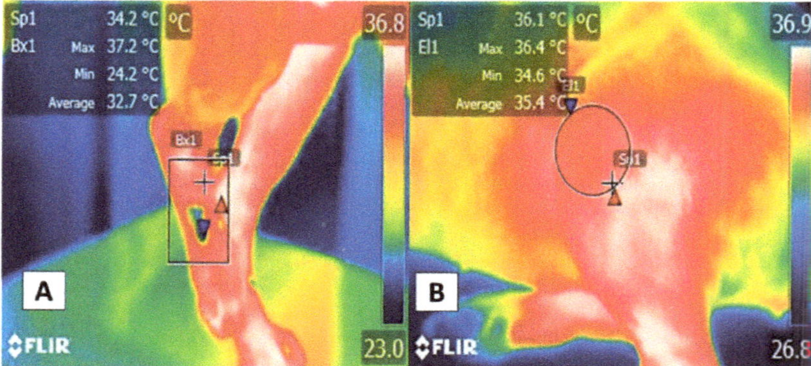

Figure 4. Modified thermal windows in forelimbs and hindlimbs for complementary diagnoses of inflammatory pathologies: (**A**) Complementary diagnosis of a metacarpal fracture. The rectangular figure (Bx1) shows a significant temperature increase in the metacarpal region of the right pelvic limb with a maximum of 37.2 °C (red triangle) and a minimum of 24.2 °C (blue triangle). This was associated with a fracture of the metacarpal bones in a 4-year-old male boxer dog. It is important to note the minimum temperature reduction, which could be related to a lack of local circulation produced by the destruction of surface blood vessels. In contrast, the maximum temperature is associated with the inflammation of peripheral tissues. (**B**) Complementary diagnosis of articular pathologies. The circular figure (El1) shows the significant increase in temperature with a maximum of 36.4 °C (red triangle) and a minimum of 34.6 °C (blue triangle) in the tibia-femoral-patellar joint of a male Pitbull dog with a cranial cruciate ligament rupture, manifesting a <3 C° limp or surrender in the left pelvic limb. Both cases show that IRT can provide a complementary, non-invasive method that aids in initial approaches to patients that have suffered trauma. Sp1: default focal point of the software. Thermal images obtained using a FLIR thermal camera.

Figure 5 shows the vascularization of the pelvic limb. This is derived from the external iliac artery, derived from the femoral artery (*femoralis*) that, at the medial level of the femoral region (*regio femoralis*), continues as the saphenous artery (*saphena*) and divides into the caudal femoral artery distal (*caudalis femoris distalis*) and in the caudal tibial (*tibialis cranialis*) and cranial (*tibialis caudalis*) artery [45].

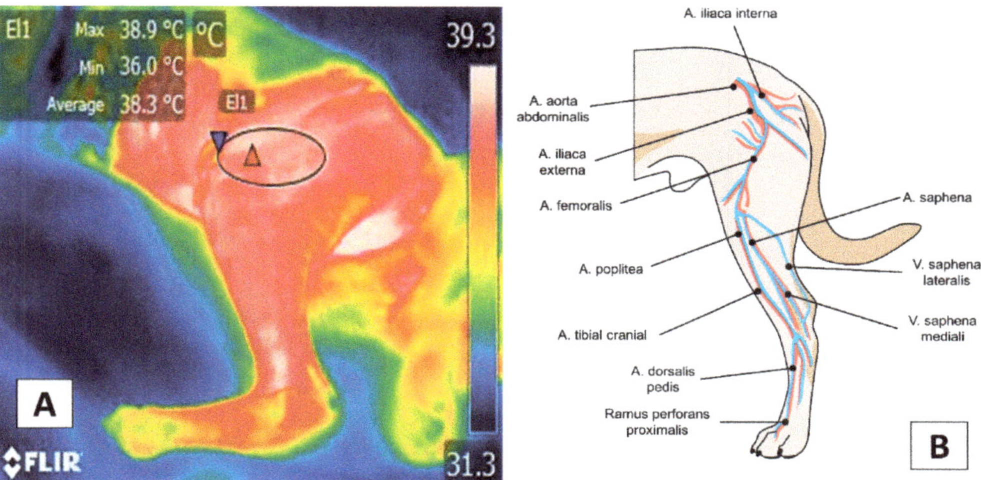

Figure 5. Appendicular window: (**A**) Femoral thermal window (El1). The tracing of this window is through an ellipse of approximately 3 cm located in the cranial region where the quadriceps femoris muscles (*quadriceps femoris*) lie and considering the cranial region of the tensor fasciae latae (*tensor fasciae latae*). (**B**) Circulation at the femoral level. The femoral thermal window obtains its circulation from the femoral artery (*arteria femoralis*) and its branch with the lateral circumflex artery (*circumflexa femoris lateralis*) that nourishes the previously described muscles that confer the importance of this window for the evaluation of the temperature. Thermal images obtained using a FLIR thermal camera.

The high density of blood vessels in the fore and hindlimbs permits visualization of the heat exchange through IRT, particularly during exercise. Rizzo et al. [73] evaluated the surface and body temperature of ten Jack Russel Terriers during 15 min walks, 10 min jogs, and 10 min gallops. After each event, the superficial temperature of the neck, flanks, ribs, back, inner thigh, and eye was collected. The results showed that the superficial temperature in the inner face of the thigh and the eye had a significant increase ($p < 0.0001$), together with increases in rectal temperature, hemoglobin levels, hematocrit, and red blood cells counts during the jogging phase ($p < 0.05$), due to the increase in blood flow in response to muscle activity. A similar event has been reported in the gastrocnemius, *biceps femoris*, and *gracilis* muscles in healthy dogs walking for 11 min. When applying a separate analysis of each muscle region, the *biceps femoris* and *gracilis* showed an increase in temperature after walking, unlike the gastrocnemius muscle; furthermore, a significant difference was detected between the temperatures of the right and left extremity [74].

Sturion [75] also applied the study of muscle groups and joints, together with cortisol levels, to determine the levels of stress and pain present in animals after performing the exercise. The author found a positive correlation between elevations in IRT, serum cortisol levels, and tympanic temperature. Therefore, detecting alterations in the muscular blood flow of broader regions such as the thoracic and pelvic limbs offers information about the muscular activity.

One study that postulates the use of appendicular thermal windows was conducted by Sung et al. [20], who evaluated 40 dogs using thermal windows in areas of the shoulder, elbow, paw, hip, knee, and hock. They were able to identify surface temperature differences between the extremities affected by bone neoplasms and reported a statistically significant difference in the average temperature between the regions of interest in their study groups (0.53 ± 0.14 °C) with a success rate of 75%–100% for identifying bone injuries.

In contrast, specific thermal windows such as those of the joints have not produced conclusive findings [15]. This was reported by Alves et al. [11], who collected 900 images of 50 work dogs, which considered the region from the last lumbar vertebra to the first coccygeal vertebra in dorsal position, while in the lateral direction, they focused on the greater trochanter region in the center. Those researchers determined that the temperature of the trochanter region was significantly higher (28.5 ± 2.8 °C) than the dorsal region (25.3 ± 9.1 °C) and mentioned a low correlation between these two regions (r^2 = 0.10, p = 0.03). These findings demonstrate that a joint such as the hip offers a thermal window that is more sensitive for diagnosing joint diseases in dogs.

Findings for cats are similar since correlation analyses of thermographic results of the region of affected tissue with the histological subtype and tumor degree in 11 cases of feline sarcoma and 31 patients with tumors in the skin or soft tissues showed that mean temperature was significantly higher in malignant tumors than benign ones (35.4 ± 01.8 °C vs. 34.4 ± 1.7 °C), and those temperatures above 34.7 °C could be associated with malignancy. That study reported a sensitivity of 76% and a specificity of 80% (p = 0.01) [76]. Studies performed to date in this species and dogs agree that IRT can be a sensitive, efficacious tool for clinically differentiating malignant from benign tumors and could aid in arriving at timely prognoses in oncological patients.

A similar response has been reported in evaluations of more specific appendicular regions, such as the gastrocnemius, *biceps femoris*, and *gracilis* muscles. For example, in healthy dogs during an 11 min walk, the analysis of individual muscles showed that the femoral biceps and *gracilis* presented temperature increases after the walk that were not evident in the gastrocnemius muscle. In addition, those authors observed a significant difference between the right and left limbs [74]. Thus, isolated muscle structures could provide an additional, more exact, window for evaluating thermal and vascular responses to exercise (Figure 6).

However, controversy has emerged because some studies report the presence of a significant influence related to the type of fur (short vs. long) that could alter IRT evaluations [77,78]. According to Nomura et al. [79], IRT was used as a complementary method in evaluating the condition of the knees of 30 dogs using the lateral, cranial, and medial faces of the joint. Those authors found no differences in the dogs' facial expressions during their assessment of this anatomical region but did observe that the temperature was 3 °C higher in the dogs with short hair than those with long hair.

This observation was reaffirmed by Kwon et al. [13] in their evaluation of the influence of the characteristics of animals' hair on the body surface, using 50 dogs acclimated to solar light for 15 min. They later obtained images of the lateral face of the femoral region and determined that the animals with a long, double coat of hair had lower temperatures (28.14 ± 0.31 °C and 28.25 ± 0.23 °C) than those that had short hair (31.77 ± 0.19 °C). Those authors concluded that this feature must be considered when evaluating appendicular extremities in this species using IRT. Something similar has been reported in horses, in which a greater density of fur was associated with a lower shoulder temperature (p < 0.001) [80].

Despite these discordant results, it has been suggested that IRT could aid in detecting temperature decreases instead of increases. This proposal is related to hemodynamic or ischemic alterations, where the literature includes reports of a negative correlation between blood pressure and average and surface temperatures based on readings from the elbow and carpal region in a porcine model [81]. Those findings indicated the need to evaluate the diagnostic precision of IRT in cats with acute paralysis to differentiate between aortic thromboembolism and non-ischemic conditions. That study reported that the mean, maximum, and minimum temperatures were lower in the right pelvic limb than on the left side. Moreover, the authors successfully identified that a temperature reduction of less than 2.4 °C in the pelvic limbs was the point for differentiating ischemia compared to controls. That work registered a sensitivity of 90% and a specificity of 100%, with positive and negative predictive values of 100% and 86%, respectively [33].

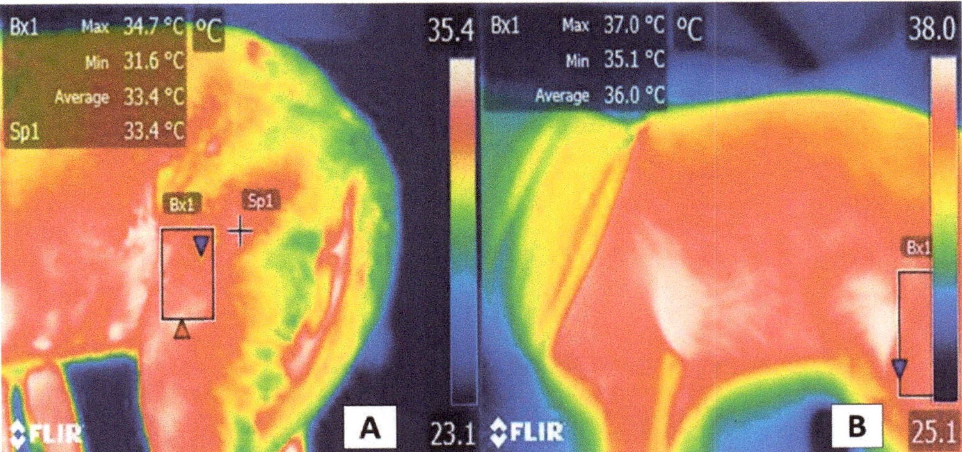

Figure 6. Thermal response of the femoral biceps muscle during moderate exercise. The thermal window of the femoral biceps muscle has been proposed to determine dogs' thermal state during exercise. This muscle spans the lateral face of the cranial zone of the femoral region, from the middle third of the femur to the proximal level of the distal epiphysis, 3 mm from the tibia-femoral-patellar joint. This zone extends to the head surface of the femoral biceps that, finally, has a tendinous insertion into the tibial tuberosity, represented by the rectangle (Bx1). (**A**) Thermal response of the femoral biceps muscle prior to exercise. This image shows the thermal response of the femoral biceps muscle (Bx1) with maximum and minimum temperatures of 34.7 °C (red triangle) and 31.6 °C (blue triangle) in a 3-year-old, non-breed female at rest. (**B**) Thermal response of the femoral biceps muscle after moderate exercise. The image shows the thermographic response after 15 min of moderate exercise during a clinical assessment of the dog's march. The significant increases of 2.3 °C (red triangle) and 3.5 °C (blue triangle) in the maximum and minimum temperatures stand out. It is suggested that this response is due to the increase in dogs' central temperature during physical activity, and hence the femoral biceps muscle receives significant vascularization from the femoral artery. Sp1: default focal point of the software. Thermal images obtained using a FLIR thermal camera.

In cats, the findings with IRT are similar to those in dogs. Quantifying the surface temperature at the limbs' levels can help detect lesions, as seen in the study by Vainionpää et al. [82], who evaluated 103 cats diagnosed with joint-level injuries. These animals underwent a physical examination, IRT valorization, and an owner questionnaire. The frequency of injuries, examined through the IRT, presented a moderate agreement of 0.48 according to a weighted kappa analysis with a confidence interval of 95%, while a low agreement was observed between owner detection of pain and thermographic images. Particularly in felines, implementing these non-invasive technologies is postulated as a valuable alternative for stress-free management during daily clinical practice [54].

Finally, the literature has suggested that these windows can help determine the degree of thermal comfort in neonatal animals. According to Reyes-Sotelo et al. [7], the low or null capacity to thermoregulate, the evaporation of the amniotic fluid, and the contact with cold surfaces are causal factors of hypothermia during this stage. In neonates, the femoral and the shoulder region are postulated as potential thermal windows [83], as indicated in humans [84].

Because of the nature of this corpus of evidence, the clinical use of appendicular thermal windows is not yet widely disseminated, so its effectiveness compared to the facial region is not clear because of characteristics such as hair type and environmental factors that reduce the reliability of these windows. Likewise, laterality probably plays a more significant role in this region since evidence indicates that the right region is more sensitive than the left. This topic must be considered in future studies.

5. Body Windows

Unlike the facial or appendicular thermal windows, thermal body windows have not been described with great exactitude, as in the case of ruminants [35,38] or pigs [85], species in which the thorax (*regio pectoralis lateralis*) or the abdomen (*regio abdominis lateralis*) are proposed as possible thermal windows to estimate the degree of comfort when exposed to extreme environmental conditions [86]. However, certain modified regions have been explored to evaluate surface lesions, as in the case of descriptions of tumors [20]. In this regard, Pavelski et al. [21] reported that the surface temperature of the mammary gland—inguinal or abdominal—was higher when affected by tumors than healthy glands. This finding suggests that IRT could function as a method for the early detection and localization of malignancy in mammary gland tumors. However, these observations await corroboration in veterinary practice, especially in clinical medicine with dogs and cats. When a malignant tumor develops, the presence of prostaglandins, serotonin, histamine, and tumor necrosis factor causes peripheral vasodilation, an event that is not observable in benign tumors [87,88]. Even though IRT is an innovative method that makes it possible to detect early changes, the lack of sensitivity and specificity studies when applied to specific body regions has reduced trust in the instrument. For this reason, one option proposed consists of complementing IRT with other clinical tools. IRT is especially sensitive to changes in the surface dermal microcirculation that occurs during skin and inflammatory diseases, and neoplasms, among other conditions [89].

The clinical application of IRT to detect malignant tumors was studied in 11 cats with feline sarcoma and 31 patients with skin or soft tissue tumors. Mean temperature was significantly higher in malignant tumors compared to benign ones (35.4 $\pm$ 01.8 °C vs. 34.4 $\pm$ 1.7 °C), and when the temperature was higher than 34.7 °C, this finding was associated with the malignancy of the tumors, with a sensitivity of 76% and a specificity of 80% (p = 0.01) [76]. In this way, both in dogs and cats, the studies agree that IRT represents a sensitive and effective tool for the clinical differentiation between malignant and benign tumors, contributing to estimating a timely prognosis in oncological patients (Figure 7).

Other studies have established the use of specific thermal windows, such as the ventral aspect of the neck. For example, Waddel et al. [90] evaluated 17 cats' hypothyroidism and 12 healthy cats by recording thermographic changes before and after cutting the animals' hair. In addition, those patients were administered radioiodine to evaluate responses at 1- and 3-month post-treatment. Authors reported an IRT precision ranging from 80.5% to 87.5% for detecting the animals with hypothyroidism before and after the hair over the ventral aspect of the neck was clipped. It might have been possible to corroborate their findings by measuring serum thyroxine levels.

These results are similar to those observed in 15 Anatolia Shepherd bitches in which thermal images were taken of the perianal and perivulvar regions to determine whether IRT could detect estrus in this species. That study determined that the surface temperature of these regions presented a non-significant increase related to serum progesterone levels and percentages of keratinized surface cells. Despite these results, the authors state that IRT could be a complementary tool for determining estrus in female dogs [91]. In addition, the authors suggest that further studies are required to determine whether IRT, as in the case of cattle [92] and pigs [93], can serve as a complementary tool to assess the reproductive status of small animals.

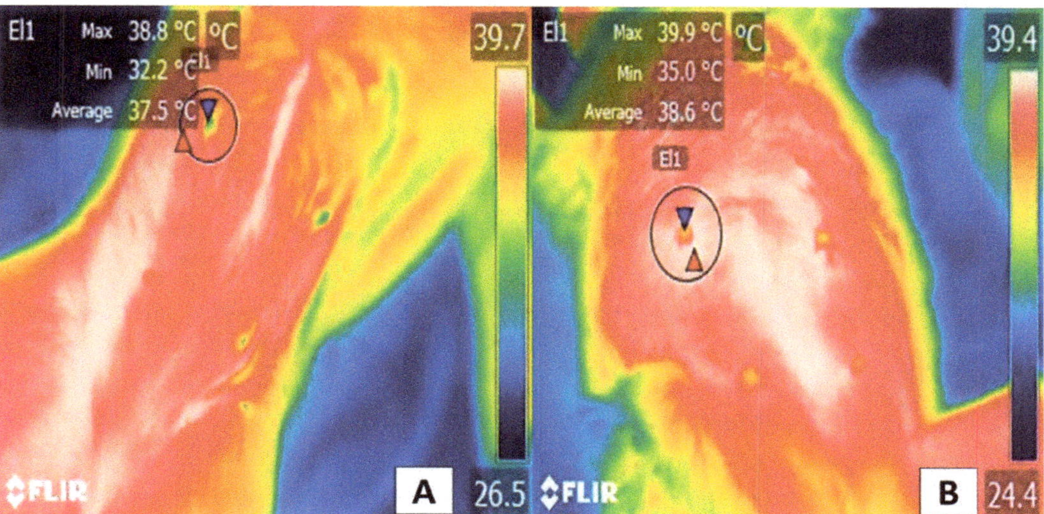

Figure 7. The thermal window of the mammary gland: (**A**) Healthy gland. The window of the mammary gland (El1) is shown in a 4-year-old Doberman Pinscher female dog. A maximum temperature of 38.8 °C (red triangle) and a minimum of 32.2 °C (blue triangle) can be seen. (**B**) Gland with the presence of tumors. An 8-year-old female dog of the Maltese breed with 0.5 to 1 cm diameter tumors in the left abdominal gland (El1). In this region, the temperature is 1.1 °C higher in the maximum values (red triangle), while the minimum is 2.8 °C higher (blue triangle), compared to the temperature shown in a healthy gland. The increase in temperature at the local level is due to cancer cells releasing pro-inflammatory substances such as serotonin, histamine, prostaglandin F2 α, and tumor necrosis factor α promoting vasodilation of superficial blood capillaries, causing an increase in the radiation emitted. Thermal images obtained using a FLIR thermal camera.

The inguinal region and the gum are other body regions studied with IRT. These have been shown to correlate moderately with the rectal temperature of 204 canines ($p < 0.001$ and $p < 0.001$, respectively), and, in particular, the gum temperature showed differences of 1 °C when compared to the rectal temperature, obtaining a sensitivity of 90% and a specificity of 78.6% to detect hyperthermia [94]. In contrast, another potential use of IRT in companion animals could be for early detection of hypothermia in areas where postural ulcers are usually present due to paralysis of the pelvic limb or severe spinal cord damage, as in human medicine [95].

This broad panorama leads to the inference that the results of these isolated studies of thermal body windows are inconclusive. Then, tasks for future research must include establishing the degree of reliability of IRT testing and determining the exact windows where specific conditions can be detected. Recent reviews suggest that windows such as the thoracic region, complemented by thoracic and pelvic limbs analyses, could be an effective method for evaluating thermal states in neonate canines, similar to findings in human medicine [7,84].

Therefore, body regions may have additional considerations for utilizing the IRT technique (e.g., type of pathology) over and above the factors mentioned about the other windows since the evidence currently available is not conclusive, even though various applications have been reported. The studies analyzed herein on body regions suggest that areas such as the neck, perivulvar, and perianal zones could be helpful to thermal windows, but it is necessary to establish their respective degrees of reliability.

6. Conclusions

The current importance of IRT demonstrates its usefulness for the early, non-invasive diagnosis of diverse inflammatory conditions, while results from some studies indicate that it could be an important method for detecting and differentiating malignant vs. benign neoplasms. Nonetheless, studies of regions such as the lacrimal caruncle, eye, or auricular region in dogs and cats have shown that these zones are especially sensitive to the activity of the ANS associated with stress factors. However, today's evidence demonstrates that the ocular region is a validated thermal window for cats with high sensitivity and specificity. For this reason, it is necessary to establish whether this is true for dogs and other potentially useful thermal windows. In addition, it is urgent to analyze biological and environmental factors concerning thermographic evaluations of different anatomical regions.

Regarding the appendicular and body thermal windows, the factors mentioned more strongly influence temperature dispersion, though one advantage is that they supply more information on the conditions of the thermal state of animals during illness. It is important to mention that using isolated regions such as the femoral biceps and *gracilis* muscles can provide more precise information on thermal states after exercise or on local inflammatory conditions in dogs. This technique has shown high sensitivity even in low circulation conditions in felines.

Finally, the body and appendicular regions discussed show the greatest contrast and discrepancies in their potential clinical use compared to facial windows. Therefore, additional research is required to establish the sensitivity and specificity of these areas for a possible future validation of the use of IRT in dogs and cats.

Author Contributions: Conceptualization, D.M.-R.; methodology, D.M.-R., J.M.-B., P.M.-M., I.H.-A., A.C.-A. and A.D.-O.; investigation, A.C.-A., A.D.-O., D.M.-R., K.L.-G. and J.G.-P.; writing—original draft preparation, D.M.-R., I.H.-A., P.M.-M., A.C.-A., J.M.-B. and P.M.-M.; writing—review and editing, D.M.-R., I.H.-A., A.D.-O., A.C.-A., P.M.-M. and J.M.-B.; supervision, D.M.-R., I.H.-A. and J.M.-B.; project administration D.M.-R., I.H.-A., P.M.-M. and J.M.-B. All authors have read and agreed to the published version of the manuscript.

Funding: This research received no external funding.

Conflicts of Interest: The authors declare no conflict of interest.

References

1. Aquino Rocha, J.H.; Vieira Póvoas Tavares, Y. Infrared thermography as a non-destructive test for the inspection of reinforced concrete bridges: A review of the state of the art. *Rev. ALCONPAT* **2017**, *7*, 200–214. [CrossRef]
2. Casas-Alvarado, A.; Mota-Rojas, D.; Hernández-Ávalos, I.; Mora-Medina, P.; Olmos-Hernández, A.; Verduzco-Mendoza, A.; Reyes-Sotelo, B.; Martínez-Burnes, J. Advances in infrared thermography: Surgical aspects, vascular changes, and pain monitoring in veterinary medicine. *J. Therm. Biol.* **2020**, *92*, 102664. [CrossRef] [PubMed]
3. Childs, C. Body temperature and clinical thermometry. In *Thermoregulation: From Basic Neuroscience to Clinical Neurology, Part II*; Romanovsky, A.A., Ed.; Elsevier: Edinburg, Scotland, 2018; pp. 467–482.
4. Lezama-García, K.; Mota-Rojas, D.; Pereira, A.M.F.; Martínez-Burnes, J.; Ghezzi, M.; Domínguez, A.; Gómez, J.; de Mira Geraldo, A.; Lendez, P.; Hernández-Ávalos, I.; et al. Transient Receptor Potential (TRP) and thermoregulation in animals: Structural biology and neurophysiological aspects. *Animals* **2022**, *12*, 106. [CrossRef] [PubMed]
5. Lush, J.; Ijichi, C. A Preliminary investigation into personality and pain in dogs. *J. Vet. Behav.* **2018**, *24*, 62–68. [CrossRef]
6. Travain, T.; Colombo, E.S.; Grandi, L.C.; Heinzl, E.; Pelosi, A.; Prato Previde, E.; Valsecchi, P. How good is this food? A study on dogs' emotional responses to a potentially pleasant event using infrared thermography. *Physiol. Behav.* **2016**, *159*, 80–87. [CrossRef] [PubMed]
7. Reyes-Sotelo, B.; Mota-Rojas, D.; Martínez-Burnes, J.; Olmos-Hernández, A.; Hernández-Ávalos, I.; José, N.; Casas-Alvarado, A.; Gómez, J.; Mora-Medina, P. Thermal homeostasis in the newborn puppy: Behavioral and physiological responses. *J. Anim. Behav. Biometeorol.* **2021**, *9*, 1–25. [CrossRef]
8. Travain, T.; Colombo, E.S.; Heinzl, E.; Bellucci, D.; Prato Previde, E.; Valsecchi, P. Hot dogs: Thermography in the assessment of stress in dogs (*Canis familiaris*)—A pilot study. *J. Vet. Behav.* **2015**, *10*, 17–23. [CrossRef]
9. Huggins, J.; Rakobowchuk, M. Utility of lacrimal caruncle infrared thermography when monitoring alterations in autonomic activity in healthy humans. *Eur. J. Appl. Physiol.* **2019**, *119*, 531–538. [CrossRef]

10. Stewart, M.; Stookey, J.M.; Stafford, K.J.; Tucker, C.B.; Rogers, A.R.; Dowling, S.K.; Verkerk, G.A.; Schaefer, A.L.; Webster, J.R. Effects of local anesthetic and a nonsteroidal antiinflammatory drug on pain responses of dairy calves to hot-iron dehorning. *J. Dairy Sci.* **2009**, *92*, 1512–1519. [CrossRef]
11. Alves, J.C.A.; dos Santos, A.M.M.P.; Jorge, P.I.F.; Branco Lavrador, C.F.T.V.; Carreira, L.M. Thermographic imaging of police working dogs with bilateral naturally occurring hip osteoarthritis. *Acta Vet. Scand.* **2020**, *62*, 60. [CrossRef]
12. Wang, F.-K.; Shih, J.-Y.; Juan, P.-H.; Su, Y.-C.; Wang, Y.-C. Non-invasive cattle body temperature measurement using infrared thermography and auxiliary sensor. *Sensors* **2021**, *21*, 2425. [CrossRef] [PubMed]
13. Kwon, C.J.; Brundage, C.M. Quantifying body surface temperature differences in canine coat types using infrared thermography. *J. Therm. Biol.* **2019**, *82*, 18–22. [CrossRef]
14. Tattersall, G.J. Infrared thermography: A non-invasive window into thermal physiology. *Comp. Biochem. Physiol. Part A Mol. Integr. Physiol.* **2016**, *202*, 78–98. [CrossRef] [PubMed]
15. Loughin, C.A.; Marino, D.J. Evaluation of thermographic imaging of the limbs of healthy dogs. *Am. J. Vet. Res.* **2007**, *68*, 1064–1069. [CrossRef] [PubMed]
16. Wan-Tae, K.; Min-Su, K.; Sun Young, K.; Kang-Moon, S.; Tchi-Chou, N. Use of digital infrared thermography on experimental spinal cord compression in dogs. *J. Vet. Clin.* **2005**, *22*, 302–308.
17. Rekant, S.I.; Lyons, M.A.; Pacheco, J.M.; Arzt, J.; Rodriguez, L.L. Veterinary applications of infrared thermography. *Am. J. Vet. Res.* **2016**, *77*, 98–107. [CrossRef]
18. Mota-Rojas, D.; Miranda- Córtes, A.; Casas-Alvarado, A.; Mora-Medina, P.; Boscato, L.; Hernández-Ávalos, I. Neurobiology and modulation of stress- induced hyperthermia and fever in animals. *Abanico Vet.* **2021**, *11*, 1–17.
19. Schaefer, A.L.; Cook, N.; Tessaro, S.V.; Deregt, D.; Desroches, G.; Dubeski, P.L.; Tong, A.K.W.; Godson, D.L. Early detection and prediction of infection using infrared thermography. *Can. J. Anim. Sci.* **2004**, *84*, 73–80. [CrossRef]
20. Sung, J.; Loughin, C.; Marino, D.; Leyva, F.; Dewey, C.; Umbaugh, S.; Lesser, M. Medical infrared thermal imaging of canine appendicular bone neoplasia. *BMC Vet. Res.* **2019**, *15*, 430. [CrossRef]
21. Pavelski, M.; Silva, D.M.; Leite, N.C.; Junior, D.A.; de Sousa, R.; Guérios, S.D.; Dornbusch, P.T. Infrared thermography in dogs with mammary tumors and healthy dogs. *J. Vet. Intern. Med.* **2015**, *29*, 1578–1583. [CrossRef]
22. Zanghi, B.M. Eye and ear temperature using infrared thermography are related to rectal temperature in dogs at rest or with exercise. *Front. Vet. Sci.* **2016**, *3*, 111. [CrossRef] [PubMed]
23. Nääs, I.A.; Garcia, R.G.; Caldara, F.R. Infrared thermal image for assessing animal health and welfare. *J. Anim. Behav. Biometeorol.* **2014**, *2*, 66–72. [CrossRef]
24. Quimby, J.M.; Smith, M.L.; Lunn, K.F. Evaluation of the effects of hospital visit stress on physiologic parameters in the cat. *J. Feline Med. Surg.* **2011**, *13*, 733–737. [CrossRef] [PubMed]
25. Polit, M.; Rząsa, A.; Rafajłowicz, W.; Niżański, W. Infrared technology for estrous detection in *Chinchilla lanigera*. *Anim. Reprod. Sci.* **2018**, *197*, 81–86. [CrossRef] [PubMed]
26. Elias, B.; Starling, M.; Wilson, B.; McGreevy, P. Influences on infrared thermography of the canine eye in relation to the stress and arousal of racing greyhounds. *Animals* **2021**, *11*, 103. [CrossRef] [PubMed]
27. Mota-Rojas, D.; Pereira, A.M.F.; Wang, D.; Martínez-Burnes, J.; Ghezzi, M.; Hernández-Avalos, I.; Lendez, P.; Mora-Medina, P.; Casas, A.; Olmos-Hernández, A.; et al. Clinical applications and factors involved in validating thermal windows used in infrared thermography in cattle and river buffalo to assess health and productivity. *Animals* **2021**, *11*, 2247. [CrossRef] [PubMed]
28. Hildebrandt, C.; Raschner, C.; Ammer, K. An overview of recent application of medical infrared thermography in sports medicine in Austria. *Sensors* **2010**, *10*, 4700–4715. [CrossRef]
29. Van Hoogmoed, L.M.; Snyder, J.R. Use of infrared thermography to detect injections and palmar digital neurectomy in horses. *Vet. J.* **2002**, *164*, 129–141. [CrossRef]
30. Tunley, B.V.; Henson, F.M.D. Reliability and repeatability of thermographic examination and the normal thermographic image of the thoracolumbar region in the horse. *Equine Vet. J.* **2010**, *36*, 306–312. [CrossRef]
31. Mota-Rojas, D.; Olmos-Hernández, A.; Verduzco-Mendoza, A.; Lecona-Butrón, H.; Martínez-Burnes, J.; Mora-Medina, P.; Gómez-Prado, J.; Orihuela, A. Infrared thermal imaging associated with pain in laboratory animals. *Exp. Anim.* **2021**, *70*, 1–12. [CrossRef]
32. Ng, W.K.; Ng, Y.K.; Tan, Y.K. Qualitative study of sexual functioning in couples with erectile dysfunction: Prospective evaluation of the thermography diagnostic system. *J. Reprod. Med.* **2009**, *54*, 698–705. [PubMed]
33. Pouzot-Nevoret, C.; Barthélemy, A.; Goy-Thollot, I.; Boselli, E.; Cambournac, M.; Guillaumin, J.; Bonnet-Garin, J.-M.; Allaouchiche, B. Infrared thermography: A rapid and accurate technique to detect feline aortic thromboembolism. *J. Feline Med. Surg.* **2018**, *20*, 780–785. [CrossRef] [PubMed]
34. Hernández-Avalos, I.; Mota-Rojas, D.; Mendoza-Flores, J.E.; Casas-Alvarado, A.; Flores-Padilla, K.; Miranda-Cortes, A.E.; Torres-Bernal, F.; Gómez-Prado, J.; Mora-Medina, P. Nociceptive pain and anxiety in equines: Physiological and behavioral alterations. *Vet. World* **2021**, in press. [CrossRef] [PubMed]
35. Bertoni, A.; Mota-Rojas, D.; Álvarez-Macias, A.; Mora-Medina, P.; Guerrero-Legarreta, I.; Morales-Canela, A.; Gómez-Prado, J.; José-Pérez, N.; Martínez-Burnes, J. Scientific findings related to changes in vascular microcirculation using infrared thermography in the river buffalo. *J. Anim. Behav. Biometeorol.* **2020**, *8*, 288–297. [CrossRef]

36. Hankenson, F.C.; Marx, J.O.; Gordon, C.J.; David, J.M. Effects of rodent thermoregulation on animal models in the research environment. *Comp. Med.* **2018**, *68*, 425–438. [CrossRef]
37. Romanovsky, A.A. Skin temperature: Its role in thermoregulation. *Acta Physiol.* **2014**, *210*, 498–507. [CrossRef]
38. Tattersall, G.J.; Cadena, V. Insights into animal temperature adaptations revealed through thermal imaging. *Imaging Sci. J.* **2010**, *58*, 261–268. [CrossRef]
39. Seixas, A.; Ammer, K. Utility of infrared thermography when monitoring autonomic activity. *Eur. J. Appl. Physiol.* **2019**, *119*, 1455–1457. [CrossRef]
40. Sutherland, M.A.; Worth, G.M.; Dowling, S.K.; Lowe, G.L.; Cave, V.M.; Stewart, M. Evaluation of infrared thermography as a non-invasive method of measuring the autonomic nervous response in sheep. *PLoS ONE* **2020**, *15*, e0233558. [CrossRef]
41. Foster, S.; Ijichi, C. The association between infrared thermal imagery of core eye temperature, personality, age and housing in cats. *Appl. Anim. Behav. Sci.* **2017**, *189*, 79–84. [CrossRef]
42. Riemer, S.; Assis, L.; Pike, T.W.; Mills, D.S. Dynamic changes in ear temperature in relation to separation distress in dogs. *Physiol. Behav.* **2016**, *167*, 86–91. [CrossRef] [PubMed]
43. Li, D.; Menassa, C.C.; Kamat, V.R. Non-intrusive interpretation of human thermal comfort through analysis of facial infrared thermography. *Energy Build.* **2018**, *176*, 246–261. [CrossRef]
44. Evans, H.E.; de Lahunta, A. *Miller's Anatomy of the Dog*, 4th ed.; Elsevier Saunders: St. Louis, MO, USA, 2012; 849p.
45. International Committee on Veterinary Gross Anatomical Nomenclature. *Nomina Anatomica Veterinaria*; World Association of Veterinary Anatomists: Hanover, Germany, 2017; pp. 1–178.
46. Cannas, S.; Palestrini, C.; Canali, E.; Cozzi, B.; Ferri, N.; Heinzl, E.; Minero, M.; Chincarini, M.; Vignola, G.; Dalla Costa, E. Thermography as a non-invasive measure of stress and fear of humans in sheep. *Animals* **2018**, *8*, 146. [CrossRef] [PubMed]
47. Stewart, M.; Verkerk, G.A.; Stafford, K.J.; Schaefer, A.L.; Webster, J.R. Noninvasive assessment of autonomic activity for evaluation of pain in calves, using surgical castration as a model. *J. Dairy Sci.* **2010**, *93*, 3602–3609. [CrossRef]
48. Lowe, G.; McCane, B.; Sutherland, M.; Waas, J.; Schaefer, A.; Cox, N.; Stewart, M. Automated collection and analysis of infrared thermograms for measuring eye and cheek temperatures in calves. *Animals* **2020**, *10*, 292. [CrossRef]
49. Hudson, L.C.; Hamilton, W.P. *Atlas of Feline Anatomy for Veterinarians*, 2nd ed.; Teton NewMedia: Jackson, WY, USA, 2010; p. 275.
50. Całkosiński, I.; Dobrzyński, M.; Rosińczuk, J.; Dudek, K.; Chrószcz, A.; Fita, K.; Dymarek, R. The use of infrared thermography as a rapid, quantitative, and noninvasive method for evaluation of inflammation response in different anatomical regions of rats. *BioMed Res. Int.* **2015**, *2015*, 972535. [CrossRef]
51. Verduzco-Mendoza, A.; Bueno-Nava, A.; Wang, D.; Martínez-Burnes, J.; Olmos-Hernández, A.; Casas, A.; Domínguez, A.; Mota-Rojas, D. Experimental applications and factors involved in validating thermal windows using infrared thermography to assess the health and thermostability of laboratory animals. *Animals* **2021**, *11*, 3448. [CrossRef]
52. Taylor, S.; Webb, L.; Montrose, V.T.; Williams, J. The behavioral and physiological effects of dog appeasing pheromone on canine behavior during separation from the owner. *J. Vet. Behav.* **2020**, *40*, 36–42. [CrossRef]
53. Riemer, S.; Heritier, C.; Windschnurer, I.; Pratsch, L.; Arhant, C.; Affenzeller, N. A review on mitigating fear and aggression in dogs and cats in a veterinary setting. *Animals* **2021**, *11*, 158. [CrossRef]
54. Travain, T.; Valsecchi, P. Infrared thermography in the study of animals' emotional responses: A critical review. *Animals* **2021**, *11*, 2510. [CrossRef]
55. Casas-Alvarado, A. Evaluación de la Analgesia Epidural de Lidocaína Sola y en Combinación con Opioides en Perras Bajo Ooforosalpingohisterectomia Electiva Mediante Termografía Infrarroja y Respuesta Fisiometabólica Sanguínea. Master's Thesis, Universidad Autónoma Metropolitana, Ciudad de México, México, 2020.
56. Czaplik, M.; Hochhausen, N.; Dohmeier, H.; Pereira, C.B.; Rossaint, R. Development of a "thermal-associated pain index" score using infrared-thermography for objective pain assessment. In Proceedings of the 2017 39th Annual International Conference of the IEEE Engineering in Medicine and Biology Society (EMBC), Jeju, Korea, 11–15 July 2017; pp. 3831–3834. [CrossRef]
57. Antonaci, F.; Rossi, E.; Voiticovschi-Iosob, C.; Dalla Volta, G.; Marceglia, S. Frontal infrared thermography in healthy individuals and chronic migraine patients: Reliability of the method. *Cephalalgia* **2019**, *39*, 489–496. [CrossRef]
58. Mazzotti, G.A.; Boere, V. The right ear but not the left ear temperature is related to stress-induced cortisolaemia in the domestic cat (*Felis catus*). *Laterality* **2009**, *14*, 196–204. [CrossRef] [PubMed]
59. Polgár, Z.; Blackwell, E.J.; Rooney, N.J. Assessing the welfare of kennelled dogs—A review of animal-based measures. *Appl. Anim. Behav. Sci.* **2019**, *213*, 1–13. [CrossRef] [PubMed]
60. Hall, E.J.; Carter, A.J. Comparison of rectal and tympanic membrane temperature in healthy exercising dogs. *Comp. Exerc. Physiol.* **2017**, *13*, 37–44. [CrossRef]
61. Sousa, M.G.; Carareto, R.; Pereira-Junior, V.A.; Aquino, M.C. Agreement between auricular and rectal measurements of body temperature in healthy cats. *J. Feline Med. Surg.* **2013**, *15*, 275–279. [CrossRef] [PubMed]
62. Kunkle, G.A.; Nicklin, C.F.; Sullivan-Tamboe, D.L. Comparison of body temperature in cats using a veterinary infrared thermometer and a digital rectal thermometer. *J. Am. Anim. Hosp. Assoc.* **2004**, *40*, 42–46. [CrossRef]
63. Nutt, K.R.; Levy, J.K.; Tucker, S.J. A Comparison of non-contact infrared thermometry and rectal thermometry in cats. 2015, 18, 798–803. *J. Feline Med. Surg.*. [CrossRef]
64. Smith, V.A.; Lamb, V.; McBrearty, A.R. Comparison of axillary, tympanic membrane and rectal temperature measurement in cats. *J. Feline Med. Surg.* **2015**, *17*, 1028–1034. [CrossRef]

65. Saeki, K.; Kutara, K.; Iwata, E.; Miyabe, M.; Shimizu, Y.; Wada, Y.; Ohnishi, A.; Matsuda, A.; Miyama, T.S.; Asanuma, T. Noninvasive thermographic photographing as an assessment of the state of discomfort in a dog receiving radiation therapy. *Animals* **2021**, *11*, 2496. [CrossRef]
66. Mota-Rojas, D.; Marcet-Rius, M.; Ogi, A.; Hernández-Ávalos, I.; Mariti, C.; Martínez-Burnes, J.; Mora-Medina, P.; Casas, A.; Domínguez, A.; Reyes, B.; et al. Current advances in assessment of dog's emotions, facial expressions, and their use for clinical recognition of pain. *Animals* **2021**, *11*, 3334. [CrossRef]
67. Fernández-Cuevas, I.; Bouzas, J.C.; Arnáiz, J.; Gómez, P.M.; Piñonosa, S.; García-Concepción, M.Á.; Sillero-Quintana, M. Classification of factors influencing the use of infrared thermography in humans: A review. *Infrared Phys. Technol.* **2015**, *71*, 28–55. [CrossRef]
68. Loughmiller, J.A.; Spire, M.E.; Dritz, S.S.; Fenwick, B.W.; Hosni, M.H.; Hogge, S.B. Relationship between mean body surface temperature measured by use of infrared thermography and ambient temperature in clinically normal pigs and pigs inoculated with *Actinobacillus pleuropneumoniae*. *Am. J. Vet. Res.* **2001**, *62*, 676–681. [CrossRef] [PubMed]
69. Soroko, M.; Howell, K.; Dudek, K. The Effect of Ambient Temperature on Infrared Thermographic Images of Joints in the Distal Forelimbs of Healthy Racehorses. *J. Therm. Biol.* **2017**, *66*, 63–67. [CrossRef] [PubMed]
70. Church, J.S.; Hegadoren, P.R.; Paetkau, M.J.; Miller, C.C.; Regev-Shoshani, G.; Schaefer, A.L.; Schwartzkopf-Genswein, K.S. Influence of Environmental Factors on Infrared Eye Temperature Measurements in Cattle. *Res. Vet. Sci.* **2014**, *96*, 220–226. [CrossRef]
71. Malafaia, O.; Brioschi, M.L.; Aoki, S.M.S.; Dias, F.G.; Gugelmin, B.S.; Aoki, M.S.; Aoki, Y.S. Infrared imaging contribution for intestinal ischemia detection in wound healing. *Acta Cir. Bras.* **2008**, *23*, 511–519. [CrossRef]
72. Vainionpää, M. Thermographic Imaging in Cats and Dogs: Usability as a Clinical Method. Ph.D. Thesis, University of Helsinki, Helsinki, Finland, 2014.
73. Rizzo, M.; Arfuso, F.; Alberghina, D.; Giudice, E.; Gianesella, M.; Piccione, G. Monitoring changes in body surface temperature associated with treadmill exercise in dogs by use of infrared methodology. *J. Therm. Biol.* **2017**, *69*, 64–68. [CrossRef]
74. Repac, J.; Alvarez, L.X.; Lamb, K.; Gillette, R.L. Evaluation of thermographic imaging in canine hindlimb muscles after 6 min of walking—A pilot study. *Front. Vet. Sci.* **2020**, *7*, 224. [CrossRef]
75. Sturion, M.A.T. Termografia de Infravermelho na Avaliação de Cães-Guia em Treinamento. Ph.D. Thesis, Facultade de Medicina Veterinária e Zootecnia, Botucatu, Brazil, 2019. (in Portuguese).
76. Nitrini, A.G.C.; Cogliati, B.; Matera, J.M. Thermographic assessment of skin and soft tissue tumors in cats. *J. Feline Med. Surg.* **2020**, *23*, 513–518. [CrossRef]
77. Mota-Rojas, D.; Titto, C.G.; Orihuela, A.; Martínez-Burnes, J.; Gómez-Prado, J.; Torres-Bernal, F.; Flores-Padilla, K.; Carvajal-de la Fuente, V.; Wang, D. Physiological and behavioral mechanisms of thermoregulation in mammals. *Animals* **2021**, *11*, 1733. [CrossRef]
78. Mota-Rojas, D.; Titto, C.G.; de Mira, A.; Martínez-Burnes, J.; Gómez, J.; Hernández-Ávalos, I.; Casas, A.; Domínguez, A.; José, N.; Bertoni, A.; et al. Efficacy and function of feathers, hair, and glabrous skin in the thermoregulation strategies of domestic animals. *Animals* **2021**, *11*, 3472. [CrossRef]
79. Nomura, R.H.C.; de Freitas, I.B.; Guedes, R.L.; Araújo, F.F.; Mafra, A.C.D.N.; Ibañez, J.F.; Dornbusch, P.T. Thermographic images from healthy knees between dogs with long and short hair. *Ciência Rural* **2018**, *48*, 1–7. [CrossRef]
80. Meisfjord Jørgensen, G.H.; Mejdell, C.M.; Bøe, K.E. Effects of hair coat characteristics on radiant surface temperature in horses. *J. Therm. Biol.* **2020**, *87*, 102474. [CrossRef] [PubMed]
81. Magnin, M.; Junot, S.; Cardinali, M.; Ayoub, J.Y.; Paquet, C.; Louzier, V.; Garin, J.M.B.; Allaouchiche, B. Use of infrared thermography to detect early alterations of peripheral perfusion: Evaluation in a porcine model. *Biomed. Opt. Express* **2020**, *11*, 2431. [CrossRef] [PubMed]
82. Vainionpää, M.H.; Raekallio, M.R.; Junnila, J.J.; Hielm-Björkman, A.K.; Snellman, M.P.; Vainio, O.M. A comparison of thermographic imaging, physical examination and modified questionnaire as an instrument to assess painful conditions in cats. *J. Feline Med. Surg.* **2013**, *15*, 124–131. [CrossRef]
83. Reyes-Sotelo, B.; Mota-Rojas, D.; Mora-Medina, P.; Ogi, A.; Mariti, C.; Olmos-Hernandez, A.; Martínez, J.; Hernández, I.; Jose, S.; Gazzano, A. Blood biomarker profile alterations in newborn canines: Effect of the mother´s weight. *Animals* **2021**, *11*, 2307. [CrossRef]
84. Topalidou, A.; Ali, N.; Sekulic, S.; Downe, S. Thermal imaging applications in neonatal care: A scoping review. *BMC Pregnancy Childbirth* **2019**, *19*, 381. [CrossRef]
85. Soerensen, D.D.; Pedersen, L.J. Infrared skin temperature measurements for monitoring health in pigs: A review. *Acta Vet. Scand.* **2015**, *57*, 5. [CrossRef]
86. Mota-Rojas, D.; Napolitano, F.; Braghieri, A.; Guerrero-Legarreta, I.; Bertoni, A.; Martínez-Burnes, J.; Cruz-Monterrosa, R.; Gómez, J.; Ramírez-Bribiesca, E.; Barrios-García, H.; et al. Thermal biology in river buffalo in the humid tropics: Neurophysiological and behavioral responses assessed by infrared thermography. *J. Anim. Behav. Biometeorol.* **2021**, *9*, 1–12. [CrossRef]
87. Axiak-Bechtel, S.M.; Mathew, L.M.; Amorim, J.R.; DeClue, A.E. Dogs with osteosarcoma have altered pro- and anti-inflammatory cytokine profiles. *Vet. Med. Sci.* **2019**, *5*, 485–493. [CrossRef]
88. Carvalho, M.I.; Silva-Carvalho, R.; Pires, I.; Prada, J.; Bianchini, R.; Jensen-Jarolim, E.; Queiroga, F.L. A comparative approach of tumor-associated inflammation in mammary cancer between humans and dogs. *Biomed. Res. Int.* **2016**, *2016*, 4917387. [CrossRef]

89. Gurjarpadhye, A.A.; Parekh, M.B.; Dubnika, A.; Rajadas, J.; Inayathullah, M. Infrared imaging tools for diagnostic applications in dermatology. *SM J. Clin. Med. Imaging* **2015**, *1*, 1–5. [PubMed]
90. Waddell, R.E.; Marino, D.J.; Loughin, C.A.; Tumulty, J.W.; Dewey, C.W.; Sackman, J. Medical infrared thermal imaging of cats with hyperthyroidism. *Am. J. Vet. Res.* **2015**, *76*, 53–59. [CrossRef] [PubMed]
91. Olğaç, K.T.; Akçay, E.; Çil, B.; Uçar, B.M.; Daşkın, A. The use of infrared thermography to detect the stages of estrus cycle and ovulation time in anatolian shepherd dogs. *J. Anim. Sci. Technol.* **2017**, *59*, 21. [CrossRef] [PubMed]
92. Talukder, S.; Thomson, P.C.; Kerrisk, K.L.; Clark, C.E.F.; Celi, P. Evaluation of infrared thermography body temperature and collar-mounted accelerometer and acoustic technology for predicting time of ovulation of cows in a pasture-based system. *Theriogenology* **2015**, *83*, 739–748. [CrossRef]
93. Simões, V.G.; Lyazrhi, F.; Picard-Hagen, N.; Gayrard, V.; Martineau, G.-P.; Waret-Szkuta, A. Variations in the vulvar temperature of sows during proestrus and estrus as determined by infrared thermography and its relation to ovulation. *Theriogenology* **2014**, *82*, 1080–1085. [CrossRef]
94. Cugmas, B.; Šušterič, P.; Gorenjec, N.R.; Plavec, T. Comparison between rectal and body surface temperature in dogs by the calibrated infrared thermometer. *Vet. Anim. Sci.* **2020**, *9*, 100120. [CrossRef]
95. Aloweni, F.A.B.; Ang, S.Y.; Chang, Y.Y.; Ng, X.P.; Teo, K.Y.; Choh, A.C.L.; Goh, I.H.Q.; Lim, S.H. Evaluation of infrared technology to detect category I and suspected deep tissue injury in hospitalised patients. *J. Wound Care* **2019**, *28*, S9–S16. [CrossRef]

MDPI
St. Alban-Anlage 66
4052 Basel
Switzerland
Tel. +41 61 683 77 34
Fax +41 61 302 89 18
www.mdpi.com

Animals Editorial Office
E-mail: animals@mdpi.com
www.mdpi.com/journal/animals